On the Reappraisal of Microeconomics

On the Reappraisal of Microeconomics

Economic Growth and Change in a Material World

Robert U. Ayres

Novartis Professor Emeritus of Management and the Environment, INSEAD, France, and Institute Scholar, International Institute for Applied Systems Analysis (IIASA), Austria

Katalin Martinás

Professor of Physics, Roland Eötvös University (ELTE), Budapest, Hungary

Edward Elgar
Cheltenham, UK • Northampton, MA, USA

Published by
Edward Elgar Publishing Limited
Glensanda House
Montpellier Parade
Cheltenham
Glos GL50 1UA
UK

Edward Elgar Publishing, Inc.
136 West Street
Suite 202
Northampton
Massachusetts 01060
USA

A catalogue record for this book
is available from the British Library

ISBN 1 84542 272 4

Typeset by Manton Typesetters, Louth, Lincolnshire, UK
Printed and bound in Great Britain by MPG Books Ltd, Bodmin, Cornwall

Contents

Figures

Tables

Acknowledgments

It is difficult to account for all of the intellectual debts we owe to many people, over a long period of time (since the late 1980s). Unfortunately some of them will inevitably be overlooked in any such attempt as this. To any such we apologize. Speaking for myself, my primary intellectual debt must be to the late Allen Kneese, who first introduced me to economic theory at Resources for the Future, back in the late 1960s, especially 1968. It was then that I became convinced that the laws of physics, notably the first and second laws of thermodynamics, had to be incorporated explicitly into economic theory. It was somewhat later that I became aware of the work of Nicolas Georgescu-Roegen and Herman Daly, with which I did not always agree, but which stimulated this continuing interest on my part. Others who contributed to my informal education back then included Ralph d'Arge, Thomas Crocker, John Krutilla, Ronald Ridker, Adam Rose and Kerry Smith. More recently I have learned a great deal from younger colleagues, especially Christian Azar, Jeroen van den Bergh, John Gowdy, Jean-Charles Hourcade, and Michael Toman. During the past fifteen years my knowledge of modern thermodynamics, which was modest to minimal when I left graduate school, owes a great deal to Stephen Berry, and the pioneering work of Ilya Prigogine and his colleagues. Later I came across the work (but not the person) of Jan Szargut, and still later that of Göran Wall, Erik Eriksson, Thomas Kåberger, Bengt Mansson, Enrico Sciubba and others encountered in the course of several Gordon Research Conferences and international energy conferences.

From mid 1986 until 1990 I spent two academic years and five summers at the International Institute of Applied Systems Analysis (IIASA) at Laxenburg, near Vienna. It was during that period that I first met Katalin Martinás, Erno Zalai and several of her other Hungarian colleagues in Budapest and became acquainted with her heterodox but fascinating ideas. To be honest, it has been the continuing interaction with her – not always smooth – that finally led to this book.

Along the bumpy road between 1990, when we left IIASA and returned to Pittsburgh, and the present moment I must acknowledge financial support from my endowed Sandoz – later Novartis – Chair at INSEAD (which I held from 1992 through 2000), and support from the MacArthur Foundation via the 2050 Project, as well as the European Commission during the years

1993–5, which enabled me to bring Kati Martinás to Fontainebleau for 18 months, during which we published several papers.

We also owe a common debt to Reiner Kümmel and the WE-Heraeus Foundation, for sponsoring a conference at Bad Honef, in November 2000, which brought us together once more and provided the final incentive to complete this book. We are very grateful to Ulrich Witt, Director of the Max Planck Institute for Evolutionary Economics, in Jena, for inviting us to present our ideas to a critical audience in September 2002. Finally we both acknowledge that this book (and several joint papers that preceded it) could never have been written without the invaluable assistance of my wife, Leslie Ayres, for many hours of technical computer backup, as well as emotional support.

RUA

In addition to the common debts noted above, I want to express my thanks to László Kapolyi, who oriented me to this question, to András Bródy who introduced me to economics; to the participants and organizers of the CSNSS workshops, and to my colleagues Sergey Amelkin, Zsolt Gilányi, Éva Hideg, Olga Kiss, István Kirschner, Béla Lukács, András Margittay-Beck, Dietmar Meyer, Michel Moreu, Aladár Nagy, Erzsébet Nováky, Konstantin Sajó, László Ropolyi, András Simonovits, Sorin Solomon, Anatoly Tsirlin, Erno Zalai, László Zsolnay. I acknowledge financial support from OTKA (T29542). Finally, for their patience and help, I have to thank my husband Guszti, my children Benedek and Veronika and my mother Katalin. I also wish to thank Leslie Ayres for her hospitality.

KM

Introduction

Our goal, when we began this work some 15 years ago, was not modest. We hoped to initiate a fundamental reconsideration of the science of economics as it applies to dynamic relationships involving the production and exchange of goods and services, while remaining fully consistent with physical laws, notably the laws of thermodynamics. This work was originally inspired, in part, by the path-breaking work on non-equilibrium thermodynamics, led by Nobel laureate Ilya Prigogine (who died as we were making final corrections to this manuscript) and his colleagues, especially Gregoire Nicolis, at the Free University in Brussels. Prigogine and Nicolis' work introduced the fundamental concept of 'self-organization' creating 'dissipative structures' characterized by maximum entropy production and driven by a source of free energy far from equilibrium (Nicolis and Prigogine 1977; Prigogine et al. 1972). The first application of this theory was to explain some peculiar phenomena in chemistry. Subsequently a number of other applications have emerged, especially in biology.

We believe that some of these ideas are also applicable in economics. This has led us to reconsider economics also as a system far from (thermodynamic) equilibrium, in which the self-organizing forces are driven by a high level of dissipation of natural resources and solar energy. This view is, of course, very different from the standard neoclassical theory currently in vogue, wherein physical laws, materials and energy are essentially ignored. Our work is more consistent in some respects with the ideas of Nicolas Georgescu-Roegen, although we disagree with one of his principal (and pessimistic) conclusions (Georgescu-Roegen 1971).

Another important source of inspiration has been the pioneering work of Richard Nelson and Sidney Winter, who have led the application of ideas from biological evolution to economics (Nelson and Winter 1982). Finally, one of us (RUA) has been fascinated by the ideas of Jane Jacobs in the field of morals and ethics in human history (Jacobs 1992). Jacobs is not a card-carrying academic scholar, but her ideas seem relevant anyhow. Otherwise, thanks to recent progress in the behavioral interface, and the increasing use of game theory to understand transactions in real markets, we find that we have nothing substantively new to add to standard critiques of neoclassical economics except perhaps a more explicit attempt to consider the role of H.

Economicus in relation to another species we call *H. Custodius* and the relationship of both to *H. Sapiens.*

Our original goal may seem, at first glance, only slightly less ambitious than the physicist's search for a 'theory of everything' (which is, of course, no such thing). As Mirowski has noted, economics, being modeled on physics, but far more complicated, may once upon a time have aspired to such a grandiose vision (Mirowski 1989a). But this is no longer even remotely possible. Having modified our original ambitions, what we still do aspire to do is two-fold. First, as already said, we want to integrate dynamic economics and the relevant part of physics, namely non-equilibrium thermodynamics. (In later work we might attempt to introduce some ideas from contemporary physics, including quantum mechanics, but that would be another story.) Second, we aspire to provide some explanation of how the long-term behavior of societies and nations can be reconciled with the short-term behavior of individuals and firms. In short, we want to create a new dynamic microeconomics that is consistent with both physics and macroeconomics, especially the resource and environmental branches. At the same time, we recognize the need to address economists in their own special language, while also providing needed explanations in ordinary language as much as possible.

The limitations of the standard axiomatic paradigm of microeconomics are well known. They have been pointed out many times. Nevertheless, criticisms focused on unrealistic or false assumptions were generally – until the 1980s – brushed aside on the grounds, most famously articulated by Milton Friedman, that falsifiable assumptions are not important if the theory makes correct predictions. No critical test of predictability has yet been identified in the domain of microeconomics that both falsifies the existing model and points clearly towards an alternative behavioral model. The difficulty (we suspect) is that the conventional near-equilibrium theory did not make many predictions, even qualitative ones, unambiguously enough to be tested in such a way, at least until the 1980s when game theorists and behavioralists began to get into the journals.

Recent work by a number of social scientists and some economists has demonstrated clearly that the Walrasian neoclassical model of behavior cannot explain a number of pertinent facts. As the author of a recent textbook has noted:

> In experiments and real life, people frequently are willing to reduce their own material well-being not only to improve that of others but also to penalize others who have harmed them or violated an ethical norm. These so-called *social preferences* help explain why people often cooperate toward common ends even when defection would yield higher material rewards, why incentive schemes based on self-interest sometimes backfire, and why firms do not sell jobs. (Bowles 2004)

Bowles' book emphasizes the role of institutions, a perspective we find very illuminating though it is not our own.

However, we note (also not for the first time) that the existing model of microeconomics fails absolutely on the macro-scale. In particular, a utility-maximization theory applicable to individuals or simple owner-operated firms operating *myopically* in a static equilibrium does not and cannot explain innovation-driven technical change (a disequilibrium phenomenon) or economic growth as it actually occurs. For this reason, theories of growth, as expressed in models, have generally relied on growth-in-equilibrium propelled by exogenous drivers. The contributions of the so-called 'endogenous' theorists starting in the late 1980s have created some doubts and offered some new directions, but have not yet been successfully quantified. By contrast, our approach to micro-foundations is arguably more realistic and yet sufficiently simple and powerful to provide the basis or 'micro-foundations' for a dynamic *presbyopic* theory that can deal with disequilibrium, growth and change at both the micro-scale and the macro-scale.

Our new approach focuses on material wealth rather than abstract utility. It does not assume that preferences are immutable – they may be situation dependent. It does not assume perfect information. Bounded rationality is sufficient. Our approach demands consistency with the laws of thermodynamics, namely that the mass-balance principle must be satisfied in all transactions, while physical processes are also subject to the 'second law' (increasing entropy). In short, we depend less than standard textbooks do on convenient but unjustifiable assumptions, notably utility maximization, rationality and – most of all – perpetual equilibrium. We accept the notion that economic agents will maximize something like neoclassical utility to the extent that is possible subject to other constraints, while noting that much can be explained by the less demanding behavioral rule that economic agents will *not* consistently make decisions that will reduce their material wealth. (In this sense, our theory is closely related to so-called 'rational expectations'.)

Our contributions to fundamental theory, we think, are as follows: (1) to modify and broaden the conventional utility theory, making it applicable to non-equilibrium and dynamic situations, and consistent with decision theory (2) to introduce the laws of thermodynamics explicitly into production and consumption processes without invoking an awkward and hard-to-explain symmetry between them and (3) to offer a new tool for simulation analysis, as an alternative to the standard constrained maximization approach. Finally, we hope to treat technology and technological change in a more realistic way, although that discussion is mostly reserved for a subsequent book.

For a time, having been inspired by the work of Nelson and Winter previously cited, we seriously thought of our work as a contribution to the sub-specialty known as evolutionary economics. Our theory is evolutionary in the sense that it reflects irreversibilities (of several kinds) and is consistent with a quasi-Darwinian rule of survival of the 'fittest', leaving the exact definition of fitness to contextual determination. However, we now think that evolutionary economics, as expounded by its chief proponents, depends too much on a questionable analogy with evolutionary biology. In biology, it is accepted that mutation is a random process driven by environmental influences, such as contact with mutagenic chemicals, climate change or ultra-violet radiation. Mutation generates diversity in the genome, while Darwinian selection operates to weed out the harmful mutations and propagate the beneficial ones. The observed pattern of alternation between periods of slow and rapid speciation (known in the trade as 'punctuated equilibrium') can, perhaps, be explained in terms of the slow buildup of beneficial mutations reaching a critical threshold.

However, we hold that discoveries and inventions do not occur in random fashion – unlike mutations. Moreover, radical general purpose innovations, which are the triggers of economic change, may – in some cases – be facilitated by an accumulation of small incremental improvements in other areas, but a necessary condition for radical change may not be sufficient. Great leaps forward are rare but of overwhelming importance to the economy. Yet they are not explained by ordinary economic behavior. In short, technological progress is not closely analogous to biological evolution, as we point out in a later chapter.

A final motivation for this book is that neoclassical theories and oversimplified models with unrealistic assumptions about technology are being used to provide policy advice at the highest levels of government, some of which is dangerously short-sighted and – we think – fundamentally perverse. Although this book is not policy oriented, we are uncomfortable with any theory that assumes existing trends are optimal – that this is, in effect, 'the best of all possible worlds'. We want to call attention to this fundamental problem and offer some suggestions as to viable alternative models.

Chapter 1 reviews and discusses the behavioral issues. *Chapter 2* reviews and summarizes the inadequacies of the standard neoclassical paradigm, even though they are well known. *Chapter 3* introduces our proposed alternative set of micro-foundations, namely the decision rules governing the actions of individuals and firms. *Chapter 4* develops the properties of the Z-function, which is the main analytical building block of the new theory. *Chapter 5* continues with a detailed discussion of short-term economic decision-making and markets where the role of technology (knowledge) can be neglected. *Chapter 6* extends the static theory into the time dimension, making it dynamic and enabling us to create a more realistic theory of production and trade. The importance of the Z-function here may be that it need only be differentiable to permit a dynamic theory. Integrability conditions, as in the neoclassical theory, are not required. In this chapter we also begin the discussion of the role of learning and knowledge as drivers of progress in the long term. *Chapter 7* discusses the transition from micro to macro, and the problems of aggregation. *Chapter 8* focuses on phenomena that occur only at the macro-level and that require a different and larger framework. Money plays a central role in linking the micro-world to the macro-world. At the micro-level, the quantity of money in the system is implicitly assumed to be constant, but its societal role is invisible. In the macro-world, money is an endogenous variable of the system. Similarly, technology is exogenously given at the micro-level, but is at least partly endogenous at the macro-level. But invention, diffusion and innovation are macro-phenomena that are also strongly influenced by external circumstances and events. We also consider, albeit very briefly, the problem of aggregate social wealth, and the relationship between the standard measure (GDP) and other factors.

In *Appendix A* we consider the multiple roles of money as a medium of exchange, a surrogate for wealth, and as a social artifact. *Appendices B, C,* and *D* further develop the mathematical tools and display simulation results.

1. Models of human behavior

Microeconomics deals with economic behavior at the level of individuals and firms. Consequently, it impinges on the domain of behavioral science and psychology. Yet, behavioral science is fundamentally empirical and experimental in nature, whereas microeconomics has largely neglected experimentation, preferring to imitate mathematics by deriving theories from axiomatic foundations. (The award of the 2002 Nobel Prize in economics to two experimentalists with an interest in behavior, Daniel Kahneman and Vernon Smith, may signal a welcome shift in orientation.)

Be that as it may, the bulk of microeconomic literature up to now is model-based. A well-known summary of models of human behavior identifies five different 'pure' models that have been influential among different groups (Jensen and Meckling 1994). While the list is not necessarily definitive, it is worth recapitulating briefly, as follows:

1. The Resourceful, Evaluative, Maximizing Model (REMM)
2. The neoclassical economic (Money Maximizing) model
3. The sociological (Social Victim) model
4. The psychological (Hierarchy of Needs) model
5. The political (Perfect Agent) model

REMM is certainly the most general of the five, and though listed first, we consider it last. The second (economic) model is a reduced version of REMM, in which many elements of wealth included by REMM are neglected for analytic simplicity. (In practice, only money income or monetary wealth is maximized.) In practice, the neoclassical economic model also makes other assumptions that are useful for creating elegant theoretical models but which are significant departures from reality. We discuss the standard neoclassical economic model in more detail in the next chapter.

The sociological model differs from REMM essentially in that individual evaluation (and choice) are absent or nearly so. Humans are viewed, *en masse*, as the product of a socio-cultural environment. 'Society' sets norms and imposes costs on people who violate culturally determined norms. This model, which is consistent with the notion of historical determinism based on class struggle (e.g. working class vs. property owning class) can be regarded as one of the chief intellectual foundations of Marxism-Leninism (and centralization of power in the hands of the 'proletariat'). Fortunately the associated political ideology has now been discredited in most parts of the world.

The psychological model, originally articulated by Maslow (Maslow 1943, 1970), supposes that human needs are strictly hierarchical in terms of 'prepotency', starting with physiological needs (for food, water, sleep), followed by safety, then love and finally 'self-actualization'. The latter is a catch-all that certainly includes identity, status, income, and achievement. It predicts that individuals at low levels of income will not spend money on services or goods satisfying needs at higher levels of the hierarchy. In contrast to the REMM model and neoclassical utility theory, it postulates, instead, a hierarchy of so-called *lexicographic* (non-substitutable) preferences. The critical difference with REMM (according to Jensen and Meckling) is that REMM allows for some elasticity of substitution at any income or wealth level, but makes the elasticity itself a function of wealth.

The first to introduce lexicographic preferences in economics was the Austrian (and founder of the Austrian school), Carl Menger in his *Principles of Economics*, 1871 (Menger 1981) as noted by Georgescu-Roegen (Georgescu-Roegen 1966). For instance, the modern dispute between ecological economists who advocate 'strong sustainability' (weak substitutability) and neoclassical economists who advocate 'weak sustainability' (strong substitutability) is basically about lexicographic preferences.[1]

The political model differs from REMM in that individuals are supposed to be 'agents' of institutions, who evaluate and maximize in terms of 'the public good' or the good of the nation, tribe, sect, institution or firm rather than their own personal preferences. In fact, the agent in this model is willing to sacrifice his or her personal welfare for the welfare of the principal. This phenomenon is consistent with the 'kamikazes' of World War II, and with the Tamil Tiger and Palestinian suicide bombers of today, as well as the popular conception (a few years ago) of Japanese or Korean workers, for instance. It is certainly consistent with the behavior of all kinds of fanatics, from animal rights activists to Christian or Islamic fundamentalists. The principal–agent model is also popular in some business management schools, where employees are presumed to act on behalf of the firm, which (in turn) is presumed to be an agent of the stockholders (Ross 1973; Gabel and Sinclair-Desgagne 1995, 1998). In fact, application of the neoclassical economics to firms presupposes that firms have split personalities: they behave like individual utility maximizers in some circumstances, but like agents (of the owners) in other situations.

As noted in the Preface, economic theory and teaching is finally beginning to acknowledge and take on board some of the insights of the old political-institutional model, which had been largely forgotten since the 1940s, (e.g. Ayres 1944) or the recent textbook by Bowles (Bowles 2004).

REMM is, in some sense, a compromise among the other four models. It postulates that every individual is an 'evaluator' who cares about every aspect of society and the environment and who makes decisions based on an ordered set of (non-lexicographic) preferences and tradeoffs. According to REMM, the individual is always willing to make tradeoffs between goods, and his/her preferences are transitive and unchanging, or at least very long lasting. The individual always wants more of whatever is regarded as a good; satiation is impossible. The individual is a utility maximizer, subject to constraints such as money and time budgets and the costs of information. This generalizes the economic model in that it allows non-monetary elements of wealth. Finally, the individual is 'resourceful', that is, he or she can imagine alternative circumstances and environments, and is capable of foreseeing the implications thereof. He/she is also capable of learning from experience, adapting to changes and creating new opportunities.

REMM accepts the sociological view that socially determined norms exist and that there are costs of departing from those norms, but it also allows for the possibility that individuals are not simply victims; they can and will try to change society if the incentives are strong enough. It takes from the psychological model the generalization that human needs and wants are universal and hierarchical – there is an ordering principle – but allows for some degree of substitution or tradeoff between any two goods or services. From the political model it accepts the possibility of altruistic behavior, insofar as individuals care about others and will try to consider those interests while maximizing their own wealth. However, it is difficult to accommodate certain kinds of behavior with personal wealth maximization[2].

REMM is regarded by some as the basis for a unified theory of the social sciences. More accurately, it appears to be an attempt to explain all kinds of human behavior in terms of economic concepts. There has been some success along these lines, as exemplified by the work of Nobel laureate Gary Becker (Becker 1973, 1968). We understand it in this spirit. Nevertheless, REMM can be criticized in several ways. Indeed, mainstream theorists have been among the sharpest critics.

One fundamental problem with REMM is that utility maximization is only possible or even meaningful in a static environment. It is inconsistent with learning from experience, adapting to change and creating new opportunities. It is also inconsistent with disequilibrium, since structural change is a disequilibrium phenomenon. In our model, discussed in this book, agents do not attempt static maximization. Instead, they make decisions subject to a simpler 'avoid avoidable loss' (AAL) rule, which – while banal – is non-trivial, robust and applicable in non-equilibrium conditions and changing circumstances.

Admittedly, neoclassical economics tries to avoid the difficulty by introducing a constrained maximization of a time integral of utility. Unfortunately, this approach requires that the utility function be integrable. The integrability conditions are not trivial and – upon deeper examination – are not compatible with real-world behavior. We return to this point in the next section.

A second problem with the standard near-equilibrium economic theory (i.e. the reduced version of REMM) is that stocks and flows are often – indeed, usually – confused. This is not a major problem in a static or near-static equilibrium, where aggregate stocks are unchanging and flows are constrained by physical conservation laws (even though such laws are rarely cited explicitly). However in disequilibrium situations the relationships are far more complex and – as Nicolas Georgescu-Roegen was the first to recognize clearly – it is important to maintain the distinction between stocks and flows explicitly (Georgescu-Roegen 1966).

A third problem with REMM also concerns Solow's famous 'trinity', namely 'equilibrium, greed and perfect rationality'. It is worth discussing this issue at somewhat greater length, below.

1.1   ON MAXIMIZATION

Following the successful use of minimum principles in classical mechanics by Lagrange and Hamilton, the founding fathers of modern theoretical economics, especially Walras, Jevons, Menger, Pareto and Wicksteed, chose profit and utility maximization (or cost minimization) as foundational principles for the description of economic decision-making, without fully realizing that this approach necessarily resulted in a timeless (quasi-static) mathematical economics. From its earliest days, however, the neoclassical paradigm has been subject to empirical and theoretical critiques that have questioned the legitimacy of its maximization principles. Already in 1918, Gustav Cassel wrote

> This purely formal (… utility) theory, which in no way extends our knowledge of actual processes, is in any case superfluous for the theory of price. It should further be noted that this deduction of the nature of demand from a single principle, in which so much childish pleasure has been taken, was only made possible by artificial constructions and a considerable distortion of reality. (Cassel 1932)

In fact, the existence of the utility and maximizing principles have been challenged from various points of view.

In this connection it is curious that many, if not most, microeconomic textbooks offer no serious defense of the maximization hypothesis. It is sometimes assumed to be so reasonable, or so obvious, as to need no explanation (van den Bergh et al. 2000). As a reminder, the modern theory of consumer behavior (that is, rationality) presupposes a continuum of well-ordered preferences that is transitive, reflexive, continuous and complete. These conditions are sufficient to assure the existence of a continuous utility function, which the rational consumer is assumed to try to maximize. Monotonicity, convexity and non-satiation are other conditions that many textbooks also assume.

Many of these underlying assumptions can be, and have been, challenged. There are serious doubts about substitutability, as already mentioned. There is experimental evidence that the transitivity axiom of consumer choice is not always valid, even for individuals. Consumers get satisfaction from the satisfaction of friends or relatives (altruism), or of social norms. Notions of fairness derived from social norms and, indirectly, from institutional structures, affect bargaining relationships. Individuals are less 'free-riding' than canonical theory suggests they should be. Indeed, the behavioral assumptions in neoclassical theory make poor predictions of actual human behavior in a wide variety of contexts (for example, Gintis 2000a; Henrich et al. 2001). In any case, the axioms of consumer choice apply – if at all – only to individuals or representative agents. But for groups of two or more it cannot be proved that 'more' is always preferable to 'less', in welfare terms, although that is arguably the most fundamental assumption of economics (Chipman and Moore 1976, 1978). The list of deviations from theory goes on and on (van den Bergh et al. 2000; Bowles 2004).

The most immediate logical basis for discarding the maximization principle is that profit and utility maximization demand perfect information, which implies perfect knowledge of the future. In a rapidly changing dynamic environment, the decision, the action and the result occur at different times and therefore in different environments. Maximization of an expected value at the moment of decision does not necessarily mean maximization of the outcome. This problem is 'solved' in neoclassical theory via the assumption of continuous equilibrium, which is clearly contrary to observed reality.

Another difficulty is that a human being cannot behave simultaneously as a utility maximizer and a profit maximizer (Martinás 2003). This is because production and consumption are two distinct activities. Everyman is sometimes a producer and, at other times, a consumer. True, not every consumer is a producer (many more are employees, or agents, of producers), but all individuals are consumers, part of the time. It is elementary mathematics that two different objective functions cannot be maximized simultaneously. To mention only one conflict, both production and consumption are activities that require time and attention, which is limited. Thus profit maximization and utility maximization are inherently inconsistent. Consequently a description of the real behavior of such agents would have to incorporate both dimensions simultaneously. Neoclassical theory therefore deals with two types of economic agents, discussed separately. The producers and consumers have different targets and different optimization.

Apart from time limitations and conflict of interest, maximization in the real world is simply too difficult and too data intensive. As far back as 1939 Oxford economists Hall and Hintch (Vromen 1995) investigated whether entrepreneurs did, in fact, determine price and output policy by means of a maximization principle, as assumed in standard neoclassical theory. Instead, they found (by survey) that almost all business men followed a 'full cost' pricing rule: they started with average (not marginal) direct unit cost per unit as the base, added a percentage to cover overheads (or 'indirect' cost), and made a further markup for profit (ibid.). This pyramid scheme is still embedded in accounting practice, which is why a tiny additional cost to some component of an automobile is deemed to add many times more to the total price.

In our view – expressed hereafter in detail – *Homo Sapiens* is not a perfectly rational evaluator in the REMM sense. The perfectly rational evaluator (in the REMM classification) is *Homo Economicus*, a subspecies. The species *H. Sapiens* is a cocktail of several other subspecies, including *Homo Ludens* (the child in each of us), *Homo Faber* (the maker), *Homo Custodius* (the guardian), *Homo Philosophicus* (the seeker) and others. From a slightly different perspective *H. Economicus* is essentially the servant who ensures the material well-being of *H. Sapiens*. A real person maximizes utility if, and when, possible, but she knows that her information and knowledge base is inadequate and, anyhow, maximization is usually impossible in a dynamic, changing environment: today's optimum will not be optimum tomorrow. However our economic agent has expectations regarding the future, and she rarely makes a decision that hurts herself, and if she does, she learns from the mistake and does not repeat it. We call this the AAL rule of behavior. It is a law of such general validity that it is nearly impossible to find a counterexample except among suicide bombers and other fanatic extremists. If we see somebody apparently violating the AAL rule in commercial life, we can usually infer that she knows something what we do not know. Or, perhaps, *H. Ludens* has taken momentary control.

In the interval between acceptance for publication and submission of the final manuscript, we have become aware of a more penetrating criticism. Neoclassical models of economic growth are based on ideas of change as motion in a framework combining classical Newtonian mechanics – as reformulated by Hamilton and Lagrange – with the energetics of Helmholtz and

Ostwald, and the comprehensive electromagnetic field theory of James Clerk Maxwell. Irving Fisher suggested that an individual agent in economics could be likened to a particle in physics (Fisher 1926). He extended the analogy further: an increment of a commodity corresponds to a displacement of the particle in Euclidean space, marginal utility (or disutility) corresponds to a force operating on the particle; total utility is an integral over marginal utilities, just as energy, in mechanics, is an integral over the product of force times displacement. In this scheme, the equations of motion of a particle in rational mechanics are determined by finding the stationary state of an integral over the so-called Lagrangian function of positions (displacements) and velocities, subject to the law of conservation of energy among other possible constraints. Hamilton's equations of motion, derivable from the Lagrangian, automatically take into account the energy conservation law.

In economics, a comparable sort of Lagrangian/Hamiltonian maximization formalism can also be introduced, to maximize total utility – the usual proxy is consumption – as an integral over time. Hamiltonian version can also be derived by assuming the existence of a corresponding conservation law, although what exactly is being conserved is never quite clear. Needless to say, it is sometimes difficult to interpret the 'equations of motion' obtained by this procedure. However, a deeper problem arises from the fact that total utility itself is defined as an integral over a vector field of marginal utilities, subject to integrability conditions. These conditions have been expressed in a variety of ways (e.g. as symmetry of the Slutsky or Antonelli matrices), but most textbooks find them difficult to interpret economically.

However, we cannot resist the following quote (Mirowski 1989a, pp. 370–1):

Now what is the economic meaning of the obscure, abstruse, deep, complex, opaque principle of integrability? It is simply this: The utility gradient must be a conservative vector field if it is subject to deterministic constrained maximization. This, in turn, dictates that utility must be path independent – that is, *by whatever sequence of events one arrives at a particular commodity bundle, one must always experience the identical level of utility* ... this is the principle that guarantees that there is something more than a mere preference ordering being represented by a utility function ... that, in principle ... utility ... should be measurable. Why? Because the conservation principles [*of energy or utility*] dictate that phenomenon W remains 'the same' under transformations A, B, ... ,N and therefore will report the same measurement under repeated examination ... In other words, it states that the conditions for an algebra are met by the phenomenon ... *the conditions for such an algebra are not met in the empirical world of markets and psychology.* A little bit of introspection should normally suffice to reveal the outlandish character of neoclassical preferences, but for the true believer there also exists a vast literature based on controlled experimentation that testifies to their spurious character (Shoemaker 1982; Tversky and Kahneman 1981); (emphasis added by RUA).

In short, the integrability conditions for the utility function are fairly rigorous, and appear to be inconsistent with real behavior. The need for such conditions has been argued at length in the economics literature, and the arguments pro and con cannot be further discussed here.

In this book we develop a mathematical model of economic activity in which there is no need for constrained maximization. The AAL rule is a central postulate. It yields a mathematical description of the economic agents and the economic system taken as a whole.

1.2   ON VALUES AND BARGAINING

According to REMM everyone is an evaluator with an unchanging value ordering scheme that affects every decision. However, in reality some decisions have the effect of long-term commitments that change and constrain subsequent valuation criteria. Joining a certain street gang, religious sect, club, organization or profession may be voluntary at the outset, but it can have profound long-term consequences insofar as valuation criteria – or moral systems – are concerned.

There are many examples. However, for our purposes, some of them can be grouped into two large generic moral systems, as proposed by social critic Jane Jacobs (Jacobs 1992). The Darwinian reference is obvious. Jacobs defines two moralities applicable to quite different groups of people. As will be seen, she is essentially contrasting *H. Custodius* (our term) with *H. Economicus*.

The 'guardian' morality – to use Jacobs' non-pejorative term –for Jacobs, involves the following rules of behavior:

- Do not engage in business or trade
- Utilize power to achieve objectives ('might makes right')
- Be obedient and disciplined; respect hierarchical authority
- Adhere to ancient tradition
- Be exclusive; do not fraternize with strangers and aliens
- Be absolutely loyal to the country, tribe, sect, military organization or firm to which one belongs; betrayal is the worst sin.
- Take vengeance ('An eye for an eye, a tooth for a tooth, etc.')
- Treasure honor, even unto death (the code of the military)
- Deceive enemies or opponents for the sake of the mission
- Use leisure time for sports or arts
- Be ostentatious to advertise status and wealth
- Dispense largesse (mainly to followers, not the needy)
- Show fortitude under pressure
- Be fatalistic (when your time has come, accept it)

The commercial morality involves a different and contrasting set of behavioral rules, namely

- Negotiate, avoid force and violence wherever possible
- Seek voluntary agreements
- Be honest ('honesty is the best policy')
- Collaborate willingly with strangers and aliens for commercial purposes
- Compete vigorously, but fairly
- Respect contracts (including informal ones) and the rule of law
- Be enterprising; be open to new ideas; be innovative
- Value comfort and convenience rather than ostentation
- Allow – even encourage – collegial dissent for the sake of the objective
- Be thrifty. Save and invest for productive purposes
- Be industrious and work hard. Be efficient
- Be optimistic (your future is in your own hands)

Arguably both moral systems are internally self-consistent. The first, or 'guardian' – possibly more accurately, the 'raider' or 'pirate' – morality rather accurately describes the behavioral rules of the ruling elites of the Roman Empire (and its predecessors), of King Arthur's legendary England, of feudal Europe and feudal Japan (i.e. Bushido). It certainly applies to the Church of Rome and Islam (indeed, any established hierarchy of priests with special access to the Deity). But it also applies to modern elites, royal families and aristocrats ('noblesse oblige'), spies and counterspies, professional soldiers, police, fire fighters, judges, civil servants, and public health officials. It also applies to criminal organizations, such as the Mafia. The behavioral implications for leaders of such organizations are those described famously by Niccolo Machiavelli in his advice to 'The Prince'.

The second, contrasting, moral system applies to the commercial world, and also to the academic-scientific world. However it, too has religious links, if not religious origins. The Society of Friends ('Quakers') in the seventeenth century fit this moral syndrome almost perfectly. Their primary departure from established religion (like Cathars and Albigensians, before them) was to reject the idea that priests had a special role as interpreters of the Word of God to Man. They said that every man has that of God within him and that he must be guided by 'the inner light' of conscience. Quakers put special emphasis on non-violence, voluntary agreements (by consensus) after free discussion, on honesty, fairness, thriftiness, industriousness and modesty.[3]

Quakers abhorred ostentation, avoided arts (as frivolous) and firmly rejected social hierarchies of all kinds. Their use of the so-called plain speech ('thee', in place of the plural 'you') was a rejection of verbal recognition of social distinctions Not surprisingly, as pacifists they had difficulty governing, even the Pennsylvania colony in which they were originally the majority (Pennsylvania is still known as the 'Quaker State'). Because so many career paths were closed to them (by choice), they mostly engaged themselves in farming, trade or education. Yet many Quaker merchants became wealthy. The obvious explanation is that conspicuous honesty creates trust and trust is good for business.

The second moral syndrome also fits perfectly with the world view of Adam Smith, for instance, who attributes wealth to trade, saving, investment, accumulation and 'industriousness' (Smith 1976). It is fairly obvious that Smith's beneficent 'invisible hand' can only function in a world inhabited by workers and merchants with the trader morality, and not the 'guardian' (military or mafioso) morality.

Jacobs argues that both moral syndromes are needed in a functional society, but that they cannot be mixed or combined without serious problems. The rulers and guardians cannot function, or even exist, without people who produce goods and services, since they themselves are precluded from doing so. In ancient times the ruling elites lived by conquest, theft or (once established in power) by some form of taxation levied on farmers, traders or producers. It is argued by some historians that this division of responsibilities was a sort of bargain: the feudal landlord provided military security, the peasant provided food, while the artisan provided other goods. Of course, the balance of power was one-sided, and the lower orders usually had little or no actual bargaining power. Nevertheless, each group served the interests of the other, enough to keep the feudal system more or less stable for many centuries. Nowadays, the guardians (civil servants, for the most part) depend on tax revenues.

On the other hand, commerce and industry thrive only in a stable environment where laws and rights are clearly established, theft and fraud are discouraged and punished, disputes are settled without violence and contracts are strictly but fairly enforced. Free competitive markets cannot exist for long without rules of conduct and incorruptible enforcers of those rules. Otherwise they quickly degenerate into chaos and the law of the jungle. But the people who enforce the laws – the 'guardians' to use Jacobs' term – must themselves live by the appropriate moral syndrome, which requires (among other things) that they themselves avoid 'trade' and do not become entrepreneurs. In context, this means they serve their hierarchical superiors (and, ultimately, the law) but do not sell their services to the highest bidders in the marketplace. Jacobs' quasi-Darwinian argument for this precept is that trade (i.e. selling services), by the guardian class, inevitably leads to corruption and a breakdown of the conditions under which entrepreneurial activity functions well.

The propensity of guardians, especially the armed forces, of developing countries to seize power in order to line their own pockets is sufficiently well-known to need little comment. But such countries tend to remain poor precisely because trade and industry do not thrive there. It is noteworthy that the most advanced and prosperous countries today are the ones where politics is relatively clean and corruption is most strongly discouraged. The fact that government agencies tend to be inefficient at providing services is also an observable fact. But the explanation is now clear: government agency heads are either political appointees or civil servants, but either way they are not supposed to be 'in business' and they have different incentives from businessmen. Privatization, if carried out properly, is therefore a sensible policy for some public services, such as transport and telecommunication.[4]

It is no accident that the moral code of traders and producers is entirely consistent with the behavior of *H. Economicus*, the idealized actor of neoclassical economics, which is the creed taught in most universities and textbooks. The term *H. Economicus* has been used to express the difference between this coldly calculating (evaluating) being and *Homo Emoticus*, its supposed obverse, a being who is ruled by emotions and feelings. However this dichotomy is certainly false and misleading, as well as derogatory. If Jacobs is right (and we think she is), one of the several subspecies of *H. Sapiens* is certainly *H. Economicus* and one of the others – but not the only one – might better be labeled *H. Custodius*.

Some neoclassical economists, will argue that *H. Custodius* is really just a utility maximizer under a different hat. The idea has been suggested that Palestinian suicide bombers are maximizing their expected rewards – supposedly including private access to a number of virgins – in Muslim paradise. (What does that imply for women?) Japanese kamikaze followers of the Bushido tradition were maximizing ... what? Pride? Glory? Duty? Needless to say, we think such explanations are too far-fetched to be credible. In this example we suggest that the situation is better understood in terms of the existence of *H. Custodius*, who exists to some degree alongside *H. Economicus* in every member of *H. Sapiens*.

The two species differ in many ways. In the first place, there is a whole cluster of fundamentally different attitudes involved, typically inculcated from childhood. (For instance, career soldiers often follow in the footsteps of earlier generations, as do career politicians, doctors, lawyers and teachers.) Sometimes there is no real choice at all, as when a girl child in South Asia is sold by destitute parents into prostitution or a Muslim or Hindu girl is forced by her parents into marriage with an older man who may be a complete stranger, or when a boy child is kidnaped and turned into a soldier (as recently in parts of Africa).

But even in Western countries where education and civil rights are nearly universal, some career choices are necessarily made very early, without much information about the implications, and yet with very long-lasting commitments and moral consequences. Once a soldier has taken his oath ('taken the king's shilling') he is not free to reconsider his loyalties every time an interesting new opportunity pops up. Soldiers, police and most civil servants are trained and expected to follow orders from superior officers, or (in the absence of a specific order) to follow established procedures spelled out in detail to cover every contingency. Lower ranks are not expected or free to think for themselves, and may be punished severely for doing so. Part of the reason for this 'bureaucratic' approach is precisely to prevent corruption – another term for buying and selling of favors and services – by eliminating almost all discretion on the part of functionaries.

Most professions require long and arduous training and entering one of them means embracing a code of conduct that prohibits certain kinds of entrepreneurial behavior, however profitable or pleasant in the short term. To break the code of conduct means leaving the profession. Professional ethics does not totally eliminate private enterprise, but it sharply restricts its scope. It is unethical for a doctor to neglect a poor patient in order to treat a rich one, even though it happens. There are comparable examples in most professions.

It is clear that *H. Sapiens* is always a mixture of *H. Economicus*, *H. Custodius,* and others as well, at least to the extent of adopting different modes of behavior under different circumstances. For instance, even the most ardent financial utility maximizer on Wall Street or Main Street must sometimes behave like a guardian or custodian in his (or her) role as a parent of young children, or as a citizen of a country under attack. Guardians do not bargain with their charges. On the other hand, even hereditary rulers must answer to their subjects, through a political process if not a marketplace, or they risk disaster. The feudal system was brought down, in the end, because the doctrine of 'divine right of kings' had been carried to the extreme of 'noblesse sans oblige' and 'let them eat cake'.

In this book, we do not attempt to grapple further with history or morality. However we think it important – indeed necessary – to begin with a clear acknowledgment that economics is a theory that does not, *and cannot*, explain all human behavior. Indeed, it is applicable to *H. Sapiens* only when and, to the extent that, he/she acts as *H. Economicus* and engages deliberately in economic activities, notably trade production and consumption. It is not applicable, on the whole, to activities of government. We return to this point in later chapters.

NOTES

1. In brief, most ecologists and environmentalists hold that some environmental services, such as climate stabilization, nutrient recycling and waste assimilation, are both fundamental to

sustainability of the biosphere and not substitutable by man-made products or services (Ayres 1978; Costanza 1991; Stern 1997).

2. It is currently popular to explain suicide bombers' willingness to die for a cause in terms of expectations of eternal bliss in some afterlife. However this explanation does not apply readily to Japanese Kamikaze or Bushido, given that Shinto is scarcely a religion at all, and there is no reason to believe that suicides expected some heavenly reward.

3. The resemblance of Quaker ideas to the well-known adages published by Benjamin Franklin in *Poor Richard's Almanack* (1733–58) is no accident. Franklin was born in Boston, but moved as a young man to Philadelphia, the city founded by William Penn (a Quaker) and dominated at the time by Quakers.

4. As too many recent Eastern European examples show, privatization without the right corollary institutions – especially a reasonably well-functioning and independent legal system – is nothing more than redistribution of public property to a few well-connected oligarchs, resulting in greater inequity and no resulting efficiency benefits.

2. Micro-foundations of economics

2.1 THE STANDARD NEOCLASSICAL ECONOMIC MODEL AND SOLOW'S 'TRINITY'

Economics is only a part (the most quantifiable part) of social science. Even so it is enormously complex. To theorize it is necessary to simplify. Neoclassical economics, which is the creed taught in most universities and textbooks, is a reduced version of the generic social science model, as described briefly in the Introduction. In particular, the description of 'economic man' is essentially a caricature. To be fair, this is increasingly recognized within the profession, and each of the various simplifying restrictions has been relaxed at one time or another in the various sub-disciplines of economics.

Two standard assumptions that permeate neoclassical economics are (1) declining marginal value (utility) of any form of wealth, (2) declining marginal returns to labor, or investment, as applied to any given activity. These two assumptions are almost two sides of the same coin, except that declining marginal utility is an assumption about individual human behavior while declining returns is an assumption about social processes. These characteristics are not universal, as it happens. There are important exceptions.[1] Cultural goods may provide increasing returns, for instance. We discuss non-declining (increasing) returns mainly in the context of technological processes, where the phenomenon is quite important. In fact, increasing returns, especially to knowledge, has been offered as a possible explanation of economic growth (Romer 1986, 1987; Lucas 1988).

A third assumption that underlies much of standard economic theory, but which we reject, is that demand for every good or service is *insatiable*. The implication of that assumption is that the demand curve for every good or service falls monotonically as the supply increases (which is consistent with declining marginal value) yet always remains positive. In the real world this may be true for money. But, as we argue in more than one place later, the contrary is often true for goods and services, especially environmental services. Too much of many a good thing can be a bad thing.

For many environmental services, in particular, there is an optimal level of supply. If the supply is sub-optimal, demand (and the corresponding shadow price) are positive. But, if there is an excess of supply, demand and shadow price become negative. Rainfall and warmth are obvious examples in point. Too little rain is a drought and too little warmth requires supplementary heat. Conversely, too much rain leads to floods and too much heat can be deadly to the unprepared. (During the summer of 2003, when temperatures in France reached record heights for nearly three weeks, there were approximately 13 000 excess deaths, mostly from dehydration or heat stroke.)

What is true of environmental services is also true of more conventional goods and services. Too little food is malnutrition; too much leads to obesity. A little sugar or salt add flavor to a meal. Too much of either is unhealthy. A little alcohol, in the form of wine or whiskey is relaxing; too much is depressing and leads to cirrhosis of the liver. A very similar point can be made with respect to trace minerals and vitamins that are essential to health. Yet almost all of them are toxic in excess quantities. In some cases (e.g. chromium and selenium) the difference between too little and too much is surprisingly small.

Insatiability may not even apply to money. Too little money is unquestionably deprivation; but too much can also be a heavy burden, at least to some people. However, the assumption of insatiability is not really necessary to the standard theory. It can be rejected without much damage to the structure.

There are some other common assumptions as well, including the assumption that markets are free, efficient and competitive, that market prices exist for all goods and services and that externalities are exceptional. All of these are dubious, to say the least. However, our major departures from the standard canon concern Solow's 'trinity', below.

2.2   SOLOW'S TRINITY

Nobel laureate Robert Solow has characterized – tongue at least partially in cheek, one supposes – three central 'structural' pillars of economic theory as 'greed, rationality and equilibrium'.[2] By 'greed' Solow apparently means 'selfishly purposeful behavior'. This translates into something like profit maximization (for firms) and 'utility maximization' (for individuals). We think it is important to note that there is a difference between short-term (myopic) profit maximization and long-term (presbyopic) profit maximization. In the previous chapter some of the criticisms of the neoclassical maximization assumption have been noted. We argue, in this and subsequent chapters, that such a strong assumption is not actually necessary to explain observed behavior. The less restrictive 'avoid avoidable loss' or AAL rule, in combination with other features of our theory, provides an adequate and elegant mathematical structure.

By 'rationality' Solow apparently means that market actors belonging to the subspecies *H. Economicus* understand their own preferences and those of

others and make optimal, utility maximizing decisions based on that understanding and on whatever budgetary and other constraints are applicable. It
follows, incidentally, that the theory in its purest form is timeless, and effectively static, because rationality presumes that each actor in the market can
foresee all the future consequences of each choice and take them into account. This, of course, implies that there are no unexpected new possibilities
being created as time goes on. Real people (especially members of *H.
Custodius*) often behave irrationally by Solow's criteria. The fireman who
risks his life to save a baby from a burning building and the soldier who risks
his life in combat are not maximizing personal utility by any economic
measure. As regards the drug addict, the slot machine player, the lottery
ticket buyer, or – an important person – the average individual inventor, it is
clear that conventional utility maximization can only be interpreted in terms
of instantaneous consumption (e.g. of entertainment) and is inconsistent with
wealth maximization. While the 'expected' gain for a lottery ticket may be
very great, such an expectation is irrational.

The inadequacy of the canonical models to policy in other areas, such as
social choice and benefit-cost analysis, has also been pointed out by a number
of authors in the mainstream literature (e.g. Boadway 1974; Bromley 1990;
Brekke and Howarth 2000; Ng 1997; Suzumura 1999. Also see Koning and
Jongeneel 1997; Bowles 2004; Gintis 2000b; Bowles and Gintis 2000).

Again, most economists recognize that the pure (timeless) form of the
theory is not adequate for purposes of dealing with long periods of time, or
multiple periods, as is necessary in several branches of macroeconomics,
such as monetary policy. Keynes, who recognized the difficulties of
endogenizing expectations (e.g. explaining inflation rates or interest rates)
finessed the problem by assuming that expectations are formed exogenously
outside the theory. Later, some macroeconomists switched to 'adaptive expectations', the idea being that expectations are changed in response to
observed differences between model results and reality. However, the theoretical adaptation process was far too slow and the underlying model errors
may remain.

The current theory, hailed as a breakthrough in some circles, is called
'rational expectations' (Lucas 1976; Newbery and Stiglitz 1982; Sargent and
Hansen 1981). The fundamental postulate, which is simple enough, is that
economic agents will not continue to make the same (i.e. systematic) errors
over and over again. In short, they will eventually learn from previous mistakes and develop model-consistent forecasts. This postulate seems obvious,
even banal. But it has the important practical implication that the equations
and parameters *in econometric models* must become policy dependent. This
insight has had important practical implications, especially in monetary policy
(for so-called 'Fed watchers').

We make a superficially similar assumption hereafter, in a slightly different context, namely, that economic agents can and do learn how to reduce risk, how to discount the future, how to bargain effectively, and how to avoid avoidable losses, as a condition of survival in the competitive market. The avoidance of avoidable losses (AAL) is, in fact, one of the key pillars of our approach.

By 'equilibrium', of course, Solow refers to the Walrasian theorem that a perfectly competitive free market will reach a (Pareto-optimal) stationary state in which the supply and demand of every commodity is in balance and the market for every commodity clears. Such a state is not static at the level of individual agents, since exchange transactions continue to occur, as goods are used and must be replaced. Some traders can increase their wealth, but only at the expense of others. The system as a whole is 'self-reproducing'. As it happens, maximizing (i.e. 'rational') behavior by individual agents is only possible in a static equilibrium, because it entails perfect knowledge of the future (prices, demands, etc.).

In a single-sector, single product economy – beloved by textbook writers – the need for equilibrium is not obvious. However, it is implicit in the common but casual use of income allocation theory, in equilibrium, to equate factor (capital and labor) input elasticities with income shares in the national accounts, at least in the case of the two-factor Cobb-Douglas production function. However, in a multi-sector, multi-product economy this simple trick does not work, as we have occasion to point out later.

The main reason for assuming equilibrium in the multi-product, multi-sector case is that it is mathematically convenient, in the sense that all the Hamilton-Lagrange optimization machinery so pervasive in the literature, depends on it.[3] However, this very convenience has led – perhaps unconsciously – to an unjustified, but pervasive assumption that the real economy is always in, or very close to, equilibrium (Solow 1970). There is also, possibly, an implicit assumption that a thorough understanding of the idealized economic equilibrium state will facilitate understanding of the non-equilibrium reality of growth and change, more or less in the spirit of Taylor expansions in mathematics or perturbation theory in physics. Our approach starts from the other perspective, namely that the equilibrium state can be understood best as the limiting case of a non-equilibrium system. It turns out that the technical problems are not as difficult to overcome as the conceptual gap.

Alfred Marshall approached the problem of growth in such an economy by starting from a stationary state in equilibrium and relaxing the conditions for a Walrasian stationary state, beginning with the constant supply of labor services (Marshall 1977, p. 305). By allowing a constant rate of growth in the labor supply he produced a steady-state growth path. Gustav Cassel formulated roughly the same idea, generalized to all the factors of production in

1932 (Cassel 1932, pp. 152–3). A similar model was formulated and solved mathematically by von Neumann in 1932 (von Neumann 1945); see also Dorfman et al. (Dorfman et al. 1958, p. 346). An important feature of the Walras–Cassel–von Neumann models is that *all products are made from other products produced within the system*. In other words, apart from labor and capital services, the system is closed.

Most explicit multi-sector growth models since von Neumann's time have been stationary general equilibrium models converted into growth models by introducing exogenous factor-augmenting technical change and increasing labor supply. The result is equi-proportionate growth in all sectors. Models with this convenient property (known as *homotheticity*) have fascinated a generation of theoreticians, resulting in numerous mathematical papers – mainly in the 1960s – on 'turnpike theorems' and 'golden rules' of capital accumulation. However they bear almost no resemblance to growth processes as they occur in the real world.

There are a number of other problems with the growth-in-equilibrium assumption. To begin with, since the work of Joseph Schumpeter (Schumpeter 1912, 1934), not to mention Robert Solow himself (Solow 1956, 1957) economic growth is acknowledged to be driven largely by technological progress and innovation. Yet, innovation, in the real world, implies disequilibrium, which creates economic incentives for innovation, whereas the neoclassical theory assumes growth-in-equilibrium, driven by exogenous forces.

A detailed critique of the equilibrium assumption from a variety of points of view has been provided by two Hungarians, Niki Kaldor (Kaldor 1971) and Janos Kornai (Kornai 1973). A point that seems crucial to us is that Walrasian equilibrium is clearly inconsistent with innovation. The reason, as noted above, is that innovation depends on the existence of unfilled needs or opportunities. It is also inconsistent with structural change at the macro-scale. Thus growth-in-equilibrium is a convenient mathematical artifact, but is essentially impossible in the real world.

It is equally important from our perspective to note that a closed system in the Walras–Cassell–von Neumann tradition has no material interactions with the natural world. Since all products are immaterial abstractions, made from abstract capital, abstract labor and from other (abstract) products from within the system, there is no need for raw materials extraction, nor disposal of material wastes. Furthermore, there is no need for negative prices, such as would be applicable (in principle) to activities like waste disposal and compensation for other externalities. In short, the Walras–Cassel–von Neumann models *have no need, and leave no room, for physical materials or exogenous inputs of energy (exergy)*. This assumption of closure is totally inconsistent with reality, as pointed out long ago by Menger (1871), and more recently by others (Boulding 1966; Georgescu-Roegen 1971; Ayres and

Kneese 1969; Ayres et al. 1970). Our approach, by contrast, treats the economy as a *processor* of physical materials and energy.[4]

2.3 EVOLUTIONARY ECONOMICS

So-called evolutionary economic theory emerged as a distinct branch in the 1980s (Guha 1981), although it really began with Joseph Schumpeter's early work, already cited (Schumpeter 1912). In standard neoclassical economics competition in equilibrium is a zero-sum game. It reflects only individual bargaining skill, since nothing is really changing in an equilibrium state. In Schumpeter's world evolutionary change is driven by needs and/or opportunities resulting from exhaustion of resources, discoveries of new resources or new capabilities, or changes in the socio-political environment (especially wars). Competitive advantage to innovators results mainly from barriers to competition, either by virtue of legal protections (e.g. patents), secrecy, or unshared institutional learning gained from experience by 'first movers'. Some neoclassical economists like Alchian and Friedman have argued that Schumpeterian competition, despite its dependence on imperfect information, is still consistent with profit maximization, because only utility maximizers will be 'selected' (in the Darwinian sense) by the market (Alchian 1950; Friedman 1953).

However it is not experimentally, or even theoretically, demonstrable that Darwinian evolution would only select utility maximizers. Besides, evolutionary economists like Winter have pointed out that the biological 'selection' analogy is inappropriate without an inheritance mechanism to assure sustainability of this behavior over time (Winter 1964). For instance, genetic selection is much less rigorous than the maximization principle implies (Kimura 1979). In fact, given a large population, subject to a wide variety of different environmental stresses, it would be very surprising if a single maximization principle could account for all selections, whereas a weaker principle might be more generally applicable. However, except for the notions of *bounded rationality* and 'satisficing' introduced by Herbert Simon (in contrast to 'maximizing') (Simon 1955, 1959), the key neoclassical assumptions (Solow's 'trinity') were not seriously challenged until the 1980s.

Another difference between evolutionary economics and the neoclassical mainstream is that neoclassical theory postulates 'representative' firms whereas evolutionary biology – and evolutionary economics – lays great stress on the existence of diversity. In fact, the mechanism that drives the economic system, in the evolutionary view, is a kind of conflict between diversity and selection. In biology, diversity of populations and species is assured by mutation combined with diversity of environments. In economics it is the result of diversity of talents and ideas among entrepreneurs, together with diversity of environments and external circumstances. The selection mechanism in biology is called 'survival of the fittest', although the precise meaning of 'fitness' is still very unclear. In evolutionary economics it varies somewhat from author to author. However in the best-known version of evolutionary theory, due to Nelson and Winter (Nelson and Winter 1982), selection is essentially equivalent to survival in the market.

Our version of evolutionary theory, described in subsequent chapters, is consistent with many of the basic assumptions of evolutionary economics as developed by earlier scholars (e.g. Boulding 1981; Hanusch 1988; Silverberg and Verspagen 1994a, 1996). However, it involves a more explicit rejection of certain neoclassical assumptions. In particular, the neoclassical theory postulates that every agent is confronted at all times by a continuum of possible choices, of which it has perfect knowledge, that the possibilities are unchanging, and among which the agent has timeless, unchanging preferences (based on self-interest) that enable it to maximize utility. By contrast, we explicitly assume that the number of choices available to an agent at any moment is limited (and may change), that the agent does not have perfect knowledge, and that the agent chooses among the (few) options available at each moment, not only on the basis of current satisfaction, but on the basis of expectations about future costs, prices, and technological capabilities: i.e. about an uncertain future. The agent maximizes, insofar as possible, but only based on expectations and among a small range of choices. The key applicable behavioral principle is to avoid avoidable losses (AAL), as we have repeatedly mentioned already.

2.4   STOCKS AND FLOWS OF GOODS AND SERVICES

Classical economic theory from Ricardo and Marx to Mill regarded value as a kind of 'substance' inherent in goods. Production, in classical economic theory, is the *source* of value, while trade merely *transmits* it and consumption *destroys* it. In this conceptualization, production is neatly balanced by consumption, whence it is also consistent with the mass balance principle. In some sense the classical theorists thought of consumption as the symmetric opposite of production (and *vice versa*).

The classical picture of value-as-substance is still implicit in the language of 'value-added', although the substance theory of value – which effectively equates value with costs of production, was harshly criticized and discarded long ago by the neoclassical marginalists, from Jevons, Walras and Wicksteed to the present day. The marginalists and neoclassicals treat goods as immaterial abstractions that acquire value only in relation to the wants and needs (i.e. the utility-field) of a consumer. Unfortunately, this conceptualization also destroys the classical symmetry between production and consumption, leaving production theory in a sort of theoretical limbo. Mirowski has identified no less than eight different formulations of neoclassical production theory, associated with most of the major names of economic history including Fisher, Gossen, Menger, Walras, Wicksteed, Marshall, Pareto, Leontiev, and Keynes (Mirowski 1989a, p. 285).

The most popular textbook approach, originally due to Wicksteed (1910) and W. Johnson (1913) re-introduces the formal symmetry between production and consumption by treating 'choice of technology' in the same way as utility theory treats 'choice of commodity'. This leads to the familiar textbook picture of hypothetical *technology frontiers*, rendered as functions of different combinations of the factors of production (capital and labor).[5] Perhaps for this reason there is little room in the standard neoclassical theory for inventories or stocks. Trading presupposes stocks of goods, both in storage and transit, as well as flows. Wealth similarly presupposes stocks of goods and money owned. Manufacturing requires stocks of raw materials, work-in-progress and generates stocks of finished products. Yet utility theory in its traditional form, harking back to Debreu's work (Debreu 1959), assumes that utility is created only by current consumption (i.e. flows), rather than by stocks.

We accept, hereafter, the crucial distinction between stocks ('funds') and flows, as emphasized especially by Nicolas Georgescu-Roegen (G-R) (Georgescu-Roegen 1966, 1984). In brief, G-R points out that certain factors or production ('funds') are unchanged by the production process, in contrast to other factors ('flows') that are incorporated in the product or used up and discarded in the process. Labor and capital equipment and infrastructure – accounted for in the canonical growth function approach, are both 'funds', whereas the relevant 'flows' of raw materials, energy and services are neglected in that scheme. Our model, discussed hereafter, is consistent with G-R's critique in that it emphasizes the material-energy aspects of the production process.

Our approach, anticipated by Kornai (Kornai 1973), also assumes that decisions by every economic agent depend upon inventories (stocks) of money and material goods. This, in itself, is not necessarily a contradiction with standard theory, since there may have been other attempts to construct utility functions dependent on stocks. In the Walrasian equilibrium case the distinction is not crucial because, while trader's stocks may vary in the short term, they necessarily remain constant, on average, over time. However, we explicitly allow stocks to change over longer periods, reflecting the possibility of accumulation for future use or depreciation resulting from obsolescence or changes in demand. Since economic decisions depend on stocks, the consequence is that the economy as a whole is necessarily time dependent.

To recapitulate, our starting point is to establish a wealth measure denoted Z, for each agent, the arguments of which are *stocks* of money and goods, rather than flows in the Debreu sense. This measure quantifies the subjective wealth attributable to the goods and money possessed by the agent (*Figure 2.1*).

A related topic that requires some elaboration is the relationship between services and material flows. As noted above, conventional utility is a function of consumption (i.e. flows of goods to consumers). At a deeper level, of course, economists have understood for a long time that it is not the goods *per se* that matter, but the services generated by those goods (e.g. Lancaster 1971). Food and drink are consumed in the literal sense, of course, but most tangible goods are only consumed in the metaphoric sense that their service output is exhausted by dissipation, wear and tear or perhaps by obsolescence. One of the links between economics and physics, also emphasized long ago, is the fact that physical wear and tear are direct consequences of the second law of thermodynamics, the so-called 'entropy law' (Georgescu-Roegen 1971). We will have more to say about this also, in a later chapter.

2.5 MARKETS

The concept of markets is so fundamental to economics that we are tempted to assume no explanation is needed. We know a market – a congregation of buyers and sellers of goods and services – when we see one. Yet, it is as well to emphasize yet again that free, competitive and efficient markets cannot exist for long without clear rules governing transactions and effective mechanisms for enforcement of the rules. It is easy to be misled by images of village marketplaces, where all transactions are personal, in cash, and the goods are portable. In the vast majority of modern markets, things are more complex. Rules are needed to define a valid and binding offer, an acceptance, a contract, a delivery, even a payment. Guarantees and warranties must be defined. Rules are also needed to deal with all the things that can (and do) go wrong in a transaction, ranging from departures from specifications and failures to perform as advertised, to 'acts of God' and deliberate fraud.

Regulation is also needed to ensure that competition in markets is reasonably fair. In most industrial countries, this means limiting the so-called 'market power' of large sellers or buyers, on the ground that price-fixing and market sharing agreements by monopolies and cartels are antithetical to competition.

Evidently the markets in the real world rarely satisfy these conditions fully, resulting in so-called 'market failures'.[6] Having said this, market failures resulting from imperfect competition or corrupt regulation are not discussed hereafter. However some market failures result from the nature of goods and services themselves. In particular, public goods and environmental services are often troublesome from a regulatory point of view, because the underlying goods themselves cannot be, or are not, owned by any individual economic agent.[7] This applies, for instance, to mobile wild animals and fish in the oceans, the atmosphere, the sun and moon, and so on. In other cases, such as some underground resources, flowing streams, parks, beaches, and 'common' land, ownership of the resource, or of access to it, is a question of law or custom. We neglect these details for the most part hereafter, although the subject of environmental resources arises explicitly in the final chapter of this book.

It is worth emphasizing once more that regulatory and enforcement functions normally cannot be provided *within* the same market, or market system, where the traders belong to *Homo Economicus*. This is because markets are institutions where goods and services are for sale to the highest bidder, whereas law enforcement and justice must be non-discriminatory. Its agents must be of the species *Homo Custodius* and therefore immune to commercial incentives.[8] This is the meaning of the powerful symbol of 'blind' justice. We assume, hereafter, that the markets discussed in this book (as in other economic texts) are adequately regulated.

2.6 RATIONALITY

It goes without saying that it is irrational to accept an offer when a better one can be achieved with little effort. Although we explicitly reject the universal applicability of *satiation* (especially in regard to environmental attributes), we acknowledge that for the majority of economic goods and services – and especially for money itself – more is normally preferable to less. The utility maximization axiom expresses this notion, albeit in an extreme form.

On the other hand, it is important to emphasize that utility maximization is *not* a necessary prerequisite for rational decisions, or for exchanges to occur and markets to exist. The minimum condition is that goods (or services) are offered by a supplier at some price, and would-be consumers are able to decide whether or not to accept the offer or make a counter-offer. There is no requirement that buyers and sellers seek and compare all possible bids, before making a choice, as would be required if the agents were true utility maximizers.

As formulated by Menger long ago, the minimum condition for an exchange is that the agents expect to be better off, and not worse off, after the exchange. This is a simple version of what we now call the 'avoid avoidable loss' (AAL) rule, which is defined more precisely later. The AAL criterion is obviously similar to the notions of *bounded rationality* and 'satisficing' (Simon 1955, 1959).

It is important to emphasize that it does *not* follow that actors will undertake any given action that satisfies the rule. It does not even guarantee that losses will not occur, from time to time, because of faulty expectations. We note, incidentally, that a utility-maximizing exchange always satisfies the AAL rule, although the contrary is not true.

From a superficial point of view, it appears that utility maximization (UM) leads to a unique outcome, whereas the AAL rule merely forbids certain possibilities. In effect, the AAL rule is sufficient, by itself, to assure directionality, on average, to economic dynamics. Putting it another way, the AAL rule assures that economic evolution must be irreversible. The choice among allowed possibilities is determined by other factors, summarized as the so-called 'force law' (derived later). The force law also incorporates the possibility of maximization, but it is not restricted to myopic maximization.

However it is important to note here that, in a bargaining situation, UM does not necessarily lead to unique outcomes any more than the AAL rule does. A simple example (below) makes the point. Here we assume that utility is a function of stocks of two goods, apples and oranges, and (implicitly) that utility is due to the consumption thereof. Based on an idea from Edgeworth, let us suppose that there are two children, Jack and Jill (Edgeworth 1925). Jack has 10 apples while Jill has 10 oranges. Assume for simplicity that they each have the same utility function: $U(\text{Jack}) = U(\text{Jill}) = A(20 - A) + O(20 - O)$ where A is the number of apples and O is the number of oranges (Martinás 1989). In the initial state the value of the utility function of each agent is 100. In other possible states we have:

Jack: 7 apples and 7 oranges, then $U(\text{Jack}) = 182$, $U(\text{Jill}) = 102$
Jack: 6 apples and 6 oranges, then $U(\text{Jack}) = 168$, $U(\text{Jill}) = 128$
Jack: 5 apples and 5 oranges, then $U(\text{Jack}) = 150$, $U(\text{Jill}) = 150$
Jack: 4 apples and 4 oranges, then $U(\text{Jack}) = 128$, $U(\text{Jill}) = 168$
Jack: 3 apples and 3 oranges, then $U(\text{Jack}) = 102$, $U(\text{Jill}) = 182$

The range of possibilities is illustrated in *Figure 2.1*. Evidently, the total utility of Jack and Jill taken together differs from case to case. The question naturally arises: Is there any necessary connection between the best possible outcome for each individual and the best possible outcome for society as a whole? This question can be formulated more precisely by introducing the concept of *Pareto-optimality*. This is the outcome of a multi-agent bargaining process *such that no agent can become better off without making another agent worse off*. It is generally assumed that such a state is not only possible, but desirable.

The initial state of Jack and Jill is not Pareto-optimal, since if Jack exchanges apples for oranges both his wealth and Jill's wealth will increase.

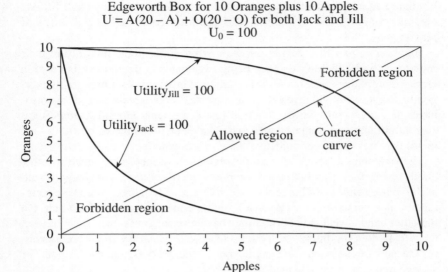

Figure 2.1   *Range of utility function possibilities*

Nevertheless there are still several Pareto-optimal final states in the above example. The Jack and Jill example also make it clear that private utility maximization, for an individual or a firm, is *not* a necessary condition for social utility maximization. Jack will maximize his personal utility if he is able to capture nearly all of the gain from a hypothetical transaction. But this can only occur if Jill accepts a nearly no-gain outcome. And, by the same token, Jill maximizes her gains if and only if Jack is willing to settle for a minimal gain. But if Jack and Jill both know the score, neither is likely to settle for a pittance, which means they will have to compromise. Evidently utility maximization for an individual in a multi-agent situation is seldom, if ever, realizable in practice since it depends upon the outcome of bargaining with other players (see next section).

An additional rule of some sort must therefore be introduced to select among the possible Pareto-optimal outcomes and (if possible) obtain a unique best of all possible solutions to the social maximization problem. (We do *not* necessarily assume that, in general, social utility is the sum of the private utilities of all the agents. The present example is intended to illustrate a different point.) As it happens, one way to do this is to assume (as we did) that Jack and Jill have exactly the same individual utility and wealth functions (the two agents being identical except in name), and to impose an exogenous 'fairness' condition. In this case the benefits of any exchange will be shared equally between them. Under this condition both Jack and Jill end up with 5 apples and 5 oranges, and the total combined social utility is maximized at 300.

However in real life, people differ in many dimensions, not only in their preference tradeoffs for the goods being exchanged in the market, but also in terms of individual preferences for bundles of other market (and non-market) goods and services. So, even if we assume that both agents – or all agents – have the same wealth/utility function (as in the Jack and Jill example) the *social* optimum may sometimes be achieved by an outcome where either Jack or Jill captures the larger objective share of the benefits.

The outcome, in terms of who possesses (consumes) how much of what goods or services, clearly depends on how effectively the agents bargain with each other. A very good bargainer (i.e. a very greedy person) may be characterized as a person with a very strong preference for 'winning' as such. On the other hand a weaker bargainer may have a strong preference for 'fairness' or perhaps just 'harmony' or 'peace and quiet'.[9] However these preferences on the part of the agents are applicable to the process, not to the good or service that is being exchanged.

To conclude, the standard UM approach, with its (misleading) implication of unique outcomes, does not allow for bargaining, hence misses a major aspect of rationality. In our alternative approach, based on the AAL rule, bargaining and compromise are not only possible but (as discussed later) explicitly taken into account.

2.7   GAMES AND BARGAINING

As the Jack and Jill example illustrates, there are many possible outcomes of a binary bargaining situation that may satisfy the AAL condition. The actual outcome depends in part on individual preferences or needs for the goods available. But the outcome also depends on the relative bargaining skills and tactics of the agents. Evidently perfect rationality and utility maximization are not as widely applicable as neoclassical textbooks tend to assume.

Since the path-breaking 1944 book *Games and Economic Behavior* by von Neumann and Morgenstern (von Neumann and Morgenstern 1944), the bargaining process itself has become an important strand of economic theory. In the standard neoclassical theory bargaining is implicitly based on a village market place model (one buyer at a time, face-to-face with one seller), or the even simpler 'Robinson Crusoe' model where an individual essentially bargains with nature. In the Walrasian generalization thereof, the process is implicitly assumed to be some kind of unspecified auction (many buyers, many sellers) but no actual bargaining behavior. The neoclassical model also assumes that both buyers and sellers possess perfect information about the product, and each other's preferences, and that the product is to be exchanged for money.

The insight of von Neumann and Morgenstern was that there are many ways in which transactions in the real world do *not* resemble this simple model, and that many of them can be represented (and better understood) as games. One of the most important distinctions they made was between 'zero-sum' games and positive (or negative) sum games. The former is a generalization of most simple games. If money is at stake, there are winners and losers but the total winnings and losses in a single game add up to zero. Exchange transactions are similar, in that the revenues obtained by the seller(s) are equal to the payments by the buyer(s). Generalizing further to the economy as a whole, the sum total of expenditures and revenues in the economy also adds up to zero.

However the game between Jack and Jill, above, was an illustration of a positive-sum game such that the compromise solution gives the largest combined utility, *but not the maximum possible for each player*. The compromise outcome usually involves explicit cooperation and information exchange on the part of the players. Finding this outcome is rational, if it exists. But the real world seldom, if ever, permits players to negotiate on the basis of perfect knowledge. Negotiations often involve deception and misinformation. In many multi-player games – especially real ones involving firms or nations – the players can, and do, form (and break) alliances and coalitions. Elections and international relations are often represented as multi-player games. In short, the possibilities are extremely diverse.

One of the key features of all games between intelligent agents (but not 'games' against nature) is that players may not be rational. Indeed, 'irrational' moves may occasionally lead to wins for unskilled players. In a game with possibly irrational players, no player can be absolutely sure what the other(s) will do. There are games such that the optimal (hence rational) solution is a negative utility. The famous 'prisoner's dilemma' is an example.[10] In this case it is difficult to arrive at the social optimum ('win-win') outcome because of the high risk of a worst case ('lose-lose') outcome.

In the prisoner's dilemma, and many other situations, 'rational' behavior for a player may be to avoid the worst case outcome, rather than to seek the best case outcome. The 'rational' choice of strategy in the real world (known as the Nash equilibrium) is the one leading to the best possible outcome, based on rational choices by all the players, such that once the best outcome has been determined for a given set of options, no player can improve his outcome by retroactively changing his choice (Nash 1951, 1953). In the case of the prisoner's dilemma, this optimum is the betrayal strategy. Norbert Wiener, father of cybernetics, once remarked that 'winning' in a multi-player game requires successful coalition-building (cooperation) in the first stages, followed by well-timed betrayal of the partners.[11] There is all too much evidence of this behavior in the real world.

In what sense, if any, is it meaningful to equate utility maximization with rationality? We cannot offer a simple answer. However there is one common characteristic of many situations, like prisoner's dilemma, where players may choose to accept negative outcomes in order to avoid the possibility of much worse ones: These examples involve choices and rules created by exogenous authorities, while restricting the possibilities of negotiation between the players. This is very different from the situation in the free competitive market of our dreams.

Economic life cannot be insulated from legal, political and physical constraints. Hence, even without the complications of strategic gaming, maximization is intrinsically very difficult, if not impossible. Our response is to abandon the idea of maximization altogether, except in those limited cases where it is obviously possible and appropriate. The only necessary condition for a voluntary economic action is that it satisfy the AAL rule, which was introduced briefly earlier in this chapter, and is discussed in more detail in the next chapter.

2.8   THE BEHAVIORALIST CRITIQUE AND THE ENDOWMENT EFFECT

The Nobel Prize for economic science in 2002 was awarded to Daniel Kahneman and Vernon Smith. Kahneman was one of the two main originators (with the late Abraham Tversky) of 'prospect theory' (Tversky and Kahneman 1974, 1981; Kahneman and Tversky 1982; Tversky and Kahneman 1987). This branch of social science challenges a number of the assumptions of neoclassical theory and effectively emphasizes differences between *H. Economicus* and *H. Sapiens*. As we have already suggested, the differences between real people and neoclassical abstractions are many. One more is that real people compare their relative well-being with others, not with absolutes. This phenomenon is sometimes known as 'keeping up with the Jones'. Many people fear loss more than they crave gains. Most people also make choices differently based on the way choices are presented.

Yet one sort of irrational choice is made with awesome frequency by a large fraction of humanity, namely to invest in lotteries or bet on horses, dogs or boxing matches. On average such bettors always lose. Buying lottery tickets, playing the slot machines and betting on horses is irrational. *H. Economicus* would not make such bets. But a lot of members of *H. Sapiens* people do it so regularly that casinos and gambling are a significant component of economic activity. It seems that the prospect of a large win (and the excitement of the betting process) provide enough entertainment to justify steady and predictable monetary losses.

Another major point of difference emphasized by behavioralists is the so-called 'endowment effect'. This is a way of saying that most people place extra value on what they already own. This has been offered as an explanation of such puzzling (to neoclassical theorists) phenomena as the unwillingness of first time stockholders to sell shares bought at high prices at prices less than the prices they paid. It also tends to explain the strong resistance of many small landowners and householders to accept monetary compensation at 'fair market value' for land or homes taken by public authorities for roads or other public needs.

Economists in recent years have devised a number of experiments to test the endowment hypothesis. The tests have shown, by and large, that inexperienced traders trade less often than neoclassical theory suggests that they should, whereas experienced traders – people who trade for a living – are more likely to trade than theory suggests that they should.

We note that the theory presented hereafter is consistent with the endowment effect as applied to individuals. We argue later that the propensity of an agent to buy or sell depends upon the difference between subjective or internal value and the price offered. Whereas firms and professional traders tend to equate internal value with actual (or calculated) monetary cost, individual consumers may retain, or purchase, goods that have personal or internal valuation higher than 'market' value.

2.9 IMPERFECT INFORMATION AND UNCERTAINTY

If outcomes were deterministic and instantaneous, economic life (and mathematical economics) would be simpler. But every transaction takes some time (if only for the check to clear), and some decisions, such as production or investment decisions, require a finite amount of time, from weeks to decades. This is enough time for circumstances to change significantly. Moreover, many decisions are based on probabilities, or 'guesstimates', not certainties. As Wicksteed noted long ago, the entrepreneur is really engaged in 'a series of speculative transactions based on estimates made in advance'(Wicksteed 1910, (pp. 372–3).

Such decisions involve *uncertain expectations*, which are based on *learning*. It is implicit that where there is uncertainty, expectations can be wrong, and losses can (and do) occur even when agents are attempting to follow the AAL rule. It is important to recognize that expectation values can be very different from maximum gains or minimum losses. However, it is also true that agents attempting to maximize the expectation-value of the transaction, also run much higher risks of loss.[12] Nevertheless, as long as the expectation value of the transaction is non-negative, it can be regarded as rational.

It is sometimes possible to devise strategies that maximize potential gain subject to a limited loss. To illustrate the point, the prospective investor in a lottery may multiply his original investment by a factor of thousands or even millions, but with a vanishingly small probability. Most ticket-buyers expect to (and do) lose their investment, and the expectation value of a lottery ticket is certainly negative. Clearly the lottery is not a rational strategy to increase wealth. However, we think that the AAL rule, as defined more precisely hereafter, is more nearly applicable to real economic life than any maximization rule.

2.10   EQUILIBRIUM

The third assumption in Solow's trinity (equilibrium) is – to put it crudely – an idealization bearing little resemblance to the real world except in a few very limited special cases. Walrasian equilibrium is a static economic state in which all actors (or agents, to introduce a modern jargon) have already made all possible exchanges that leave them better off than they were before. From a macroeconomic perspective, it is a stationary state in which there is a unique set of prices such that supply and demand are perfectly matched in every sector. Leon Walras postulated the existence of such a state back in 1874. The most general existence proof appeared in 1954 (Arrow and Debreu 1954).[13]

In a Walrasian equilibrium all markets 'clear' by definition. This means that there are no surpluses – *hence no stocks* – and, of course, no shortages. A shortage is inconsistent with a stationary state, of course, since if demand exceeds supply in a free competitive market prices will rise, supply will increase and demand will fall until the supply and demand are equal. Evidently there is a special case, where the supply of all goods in circulation is fixed (i.e. there is no production, consumption or depreciation) and *only* exchanges are possible. This special case is rare in real life, but it has some analytic features that are mathematically convenient. In fact, most (all?) of the proofs of the existence of Walrasian equilibrium apply only to this special case. The more general case where production and consumption and stock changes are involved has received much less attention, for good reason. We address this point later.

Most economists, including Solow himself, believe wholeheartedly that the real economy is always quite close to a Walrasian equilibrium (e.g. Solow 1970). For this reason, perhaps, there has been virtually no effort (that we are aware of) to develop measures of economic 'distance' from equilibrium.

Another reason for neglecting this issue is that to give up the equilibrium assumption would automatically invalidate the (implicit) assumption that static optimization with perfect information is tantamount to dynamic optimization. Yet these two types of optimization are not at all equivalent. It is well known that one cannot maximize two different objective functions simultaneously, which means that one cannot maximize welfare 'now' and welfare 'tomorrow' at the same time. This mathematical fact, in turn, invalidates one of the most popular tools of the modern theorist, namely 'constrained optimization' and so-called computable general equilibrium (CGE) models.

Walrasian equilibrium is obviously not the same as thermodynamic equilibrium in physics. However, in addition to the name, the two concepts have one feature in common. Thermodynamic equilibrium and Walrasian equilibrium are both stationary states in which nothing can change at the macro-level: everything that *can* happen has already happened. Thus the notion of growth-in-equilibrium, which is so popular in economic growth models of the CGE type, is essentially an oxymoron.

The non-equilibrium approach adopted hereafter does not exclude the possibility of equilibrium, of course. It merely regards equilibrium as a somewhat unlikely special case.

2.11   GROWTH AS A DISEQUILIBRIUM PROCESS

Evolutionary economics rejects much of the neoclassical paradigm, including rationality (optimization) and equilibrium. Evolutionary theory assumes bounded rationality. It also rejects the concept of the aggregate production function. Growth and change are treated as a contest between selection for survivability and increasing diversity in a heterogeneous population of stylized firms. The population is differentiated in terms of proprietary techniques (routines), sizes, profitability and allowed behaviors, including search, imitation, investment, entry and selection (disappearance) (Winter 1964; Nelson and Winter 1974, 1982).

Since the early 1980s there has been an explosion of interest in evolutionary models, mainly focusing on non-linear dynamics and innovation.[14] It must be said, however, that most of these contributions are purely theoretical. One alternative method of analysis for evolutionary theorists is computer simulation, via 'agent-based' models with the objective of reproducing plausible aggregate behaviors of macro-variables, from simple rules of micro-interaction (for example, Axtell and Epstein 1996; Axtell et al. 1999). In such cases the growth dynamics are derived from the behavior of individual agents in a large population, interacting according to fixed rules: for example, that firms with low rates of profit actively search for improved techniques by investing in

R&D, or that new capital is invested in sectors where capital is accumulating rapidly. Nelson and Winter focused on reproducing capital/labor ratios and increasing productivity over time, similar to what is observed in a real economy (Nelson and Winter 1982).

Evolutionary theory, as developed up to now by a number of authors, has some desirable features, as compared with neoclassical theories. One is the possibility of representing the phenomenon of path-dependence and 'lock-in', which is impossible in a growth-in-equilibrium theory. Another is a clearer distinction between growth and change. Equilibrium growth theory excludes structural change, whereas evolutionary theory probably allows it. In fact, evolutionary theory also permits growth without change (pure replication) or change without growth, or even negative growth.

Evolutionary theory, to date, does not explicitly include a role for knowledge accumulation, except insofar as it is equivalent to memory as applied to the search process. It does not really distinguish between knowledge associated with small and gradual improvements due to learning-by-doing or learning-by-using, *vis-a-vis* new theoretical knowledge attributable to major scientific discoveries resulting in breakthrough inventions and innovations. In short, evolutionary economics does not really distinguish between Schumpeterian and Usherian processes. Yet the distinction is vital, because the former can be extremely destructive while the latter are generally benign. Moreover, as emphasized in Chapter 8, Schumpeterian innovations are often the result of irrational risk-taking by individuals that cannot really be explained in economic terms.

Perhaps more to the point, evolutionary theory to the best of our knowledge does not attempt to explain many economic activities, including consumption. It focuses almost exclusively on change *per se*. In contrast, we attempt to introduce knowledge as a component of wealth and as a core variable to explain exchange transactions and (especially) production theory.

Our approach, described hereafter, shares many of the key features of evolutionary economics. However it permits more use of conventional analytical techniques, although by no means excluding the use of simulation as a supplementary tool. In fact, a number of simulations are presented in this book, especially in the Appendices.

NOTES

1. The inapplicability of the standard axioms of consumer choice were discussed in the previous chapter (see also Bowles 2004; Gintis 2000b; Bowles and Gintis 2000).
2. We haven't identified the original source. It probably doesn't matter.
3. Here we refer again to the standard technique of maximizing, not at a point in time, but an integral over time. In case there are any readers unfamiliar with the mathematics of utility maximization, it is important to note that there are restrictions on the functional forms that can be subjected to this treatment. These restrictions are known as 'integrability condi-

tions', meaning that neoclassical utility must be a conservative vector field (in commodity space) with the property of path-independence. This property, in turn, means that the utility of a bundle of commodities to a given consumer is permanent, unchanging and independent of the sequence of transactions by which it is achieved. This guarantees measurability. However, as many experiments have shown, path-independence is not a condition that is normally met in the real world (e.g. Tversky and Kahneman 1981).

4. For readers with a mathematical bent, we point out that the absence of a set of mathematical constraints resulting from the Walrasian market-clearing (equilibrium) assumption is compensated by a set of mass-energy balance conditions which have a comparable information content.

5. However, while it is convenient to imagine – as some textbooks do – that production is like mixing a salad with different combinations of simple ingredients, this picture is a gross misrepresentation of technology as it exists in the real world. Technology of production depends on the characteristics of real materials and the availability of useful energy. Labor is a complex input consisting of muscle-work and brain-work. Muscle-work can be (and has been) replaced in many situations by machines, which are forms of capital, but which utilize flows of energy. Brain-work can also be replaced to some extent by machines (computers), which are forms of capital, also requiring energy flows.

6. The problem of externalities has been discussed from various perspectives by a number of economists, e.g. Pigou (1952), Scitovsky (1954), Coase (1960) and others. However virtually all of the early literature approached the problem from a partial equilibrium perspective and assumed that the external effects were unusual and not pervasive. The idea that externalities are inherently pervasive and that they result from waste residuals that, in turn, are consequences of the central role of materials and energy in the economic system was first noted by Ayres and Kneese (Ayres and Kneese 1969).

7. One of the first attempts to emphasize the importance of individual ownership – or its absence – was a paper entitled 'The Tragedy of the Commons' (Hardin 1968).

8. Although it is perhaps conceivable that police and law enforcement services could be provided by competing firms, with loyal and incorruptible employees, operating in another higher level market. Such 'privatization' possibilities have become increasingly attractive to conservatives, in recent years, although it is not yet clear whether the problem of conflict of interest can be satisfactorily resolved.

9. The bargaining problem has recently attracted serious attention from experimental economists. To mention one example, Werner Güth of the Max Planck Institute devised the so-called 'ultimatum game', which captures some of the elements of the bargaining situation. In this game there are two players who cannot communicate. One is chosen at random. This player is given $100 to be shared with the other. He must propose a way to divide the money. There is only one opportunity to make an offer and one opportunity to accept or reject. If the second player accepts, the money is divided as agreed. If the other player refuses, neither player gets any money. There are no further negotiations. The game is played just once, so there can be no learning from experience. The interesting point of the game is that *Homo Economicus* would always agree to any non-negative offer, even just $1, since something is better than nothing. However real people in the responder position tend not to agree to accept less than 20% of the $100. More surprisingly, two thirds of the proposers *offer* between 40% and 50%, apparently on 'fairness' grounds.

10. For those who do not know it, two prisoners A and B, are arrested for the same crime. Unable to communicate with each other, they are faced with the following choice: if both A and B assert the other is innocent, they will both be released. If A testifies against B while B asserts that A is innocent, A is released while B gets a 10-year prison sentence, and conversely. Finally, if both A and B blame the other, both will get a 5-year term. Clearly, cooperation gives the best possible outcome for both prisoners, but since they cannot communicate, cooperation depends upon trust. On the other hand, betrayal always avoids the worst outcome and might result in the best outcome (release) if the other prisoner elects to cooperate.

11. In the chapter 'Information, language and society' of Norbert Wiener's *Cybernetics* (Wiener

1948). Hitler's non-aggression pact with Stalin in 1940, followed within months by his surprise attack on the Soviet Union, is a perfect example.

12. Normally an investor does not risk more than his actual investment, but there are investment strategies, such as short-selling shares, that can lead to much larger losses. The 'names' who underwrite insurance at Lloyd's are at risk for their entire net worth, in case of a large loss.

13. The sophisticated mathematical tools used in that proof, notably the fixed-point theorem from topology, have arguably inspired generations of theoretical economists since then to treat economics as a branch of mathematics. We are not the first to deplore that trend.

14. Major edited volumes on the topic include Day and Eliason (1986); Dosi et al. (1988); Lorenz (1989); Helmstaedter and Perlman (1996). Other pertinent papers include Day (1984, 1987); Day and Walter (1989), Dosi (1988, 1982) Silverberg et al. (Silverberg et al. 1988; Silverberg 1988; Silverberg and Lehnert 1993; Silverberg and Verspagen 1994a, 1994b, 1996); Kwasnicki (1996).

3.   Economic agents, actions and wealth

3.1   ON DECISIONS IN ECONOMICS

Our perspective can be summarized briefly. Decisions (in economic affairs) are tantamount to selections from a limited set of possibilities for immediate action. The contemplated action may or may not have long-term implications. The set of possibilities is constrained by the external environment, such as the legal framework, and by the assets of the agent. These include financial (i.e. more or less liquid) assets, physical assets and intangibles such as knowledge, know-how, reputation, and so on. The initial *state* of the agent comprises all of these conditions and assets. The criteria for selection are expected outcomes of the action.

Most important, we emphasize that for most economic agents, at most times, the scope for selection is very limited. Rather than a continuum of well-understood choices, the agent in reality is normally faced with a much simpler decision, generally based on very incomplete knowledge. Sometimes it is a simple yes/no, for example, to buy an item at the price offered, or to sell an item at the price offered. Bargaining is sometimes possible, but it usually amounts to several offers and counter-offers, each of which requires a yes or no decision, still based on limited knowledge. A given offer may or may not be the last; other offers may or may not occur. Firms must decide whether to bid on tenders, and if so how much. Factory owners or managers must decide whether to produce for a given market, or not. At most, the question is likely to be how much to produce for sale at what projected price. The only type of economic agent that really confronts a continuum of possibilities is the pure investor in stocks and shares, and there are very few such agents. In all cases, the available information is limited. It is true that a consumer with plenty of money can choose among a wide variety of ways of spending the money, but we argue (later) that consumption of this sort is not really an economic process in the strict sense.

We assume that the selection process itself is essentially finite, in the sense that there is an irreducible minimum scale beyond which an economic decision cannot be subdivided into smaller discrete decision-elements. A *unit economic process* is irreducible in that it cannot be divided into smaller process elements and still retain its economic character. For instance, if the two parts of an exchange (e.g. apples for oranges) are de-linked, they must be regarded as gifts and the avoid avoidable loss (AAL) rule does not apply.

As noted at the beginning of the last chapter, we assume declining marginal utility (admittedly not yet defined) and declining marginal returns, in concert with virtually all economists. On the other hand, as noted at that point, we do not unconditionally accept the three core but simplistic neoclassical assumptions known as 'Solow's Trinity': greed, rationality and equilibrium. We also dislike the neoclassical conceptualization of utility as an integral over a conservative vector field in commodity space – meaning that preferences are permanent and unchanging – which lies behind all the machinery of constrained optimization and the use of Hamiltonians (Mirowski 1989b). Having rejected them, however, it is necessary to provide an alternative set of structural axioms with which to construct a basic theory. Ours include the following:

1. Agents are rarely (if ever) indifferent to outcomes. Agents want to select outcomes that increase their economic welfare or state of well-being. They want to avoid losses. (On the other hand it is not assumed that more of any good is always preferable to less.)
2. Outcomes are rarely, if ever, instantaneous. They take time, sometimes (in the case of basic research or mineral resource development, real estate development or infrastructure development) they take a long time, during which circumstances can change quite a lot. Where outcomes are delayed for significant periods, the preferences that governed decisions at the commencement of the process may not be valid or applicable for later decisions, due to changes in the agents themselves, or the environment. This is one of the reasons why some losses are unavoidable.
3. Unfavorable outcomes cannot always be avoided, because even if the agent has perfect knowledge, which is rarely the case, circumstances can change between decision and outcome. The longer the delay between decision and consequence, the more likely it is that the outcome will be different from the expected outcome.
4. The difference between expectation and reality can be unfavorable in terms of the agent's values and priorities. The longer the delay between decision and outcome, the greater the probability (i.e. risk) of an unfavorable outcome and the more unfavorable the worst case outcome can be, *ceteris paribus*. This being true, rational agents seek greater payoffs when outcomes are further in the future and therefore more uncertain. This consideration implies that future expected payoffs (e.g. from investments) must be discounted in comparison with present values. (Discounting is sometimes called *time preference* for this reason.) Rational economic agents will also demand greater expected payoffs for

success in proportion to the magnitude of the possible loss, or worst case outcome. This response is called a *risk premium*.

5. Knowledge and know-how as applied to materials, products, production processes, finance, markets and the environment (legal, economic, political and cultural) are components of wealth, and multipliers of wealth. Knowledge cannot (except in a few special cases) be exchanged *as such* for goods or money, since the seller or giver does not lose what he sells or gives away. Moreover, knowledge can be transmitted only to the extent that it can be codified in language. Useful knowledge (and skill) accumulates as a result of experience and inadvertent learning, but the learning process cannot be avoided. However it can be accelerated by education or even multiplied, especially by formal research. Skills can be taught only with assistance from a lot of practice.

Rationality, for our purposes, encompasses all of the above. Evidently expectations are a crucial element in decision-making. The difference between success and failure in risky undertakings is generally attributable to an agent's skill in forming expectations and corresponding time preferences and risk premiums. These skills can be learned and (to some extent) taught. Having been learned, they constitute an element of knowledge.

Rationality, as defined above, implies definite preferences at any given moment in time, though it does not preclude changing tastes or changing preferences resulting from learning by experience. Young people like loud music, fast food, tattoos, and weird clothing fashions that older people find objectionable if not offensive. Older people have different tastes and pleasures. Similarly, groups and nations can learn from experience and change, although more slowly. As an example, the Irish potato famine in the 1840s taught the Irish people the danger of overpopulation and the absolute necessity to limit birthrates and family size. Without abandoning Catholic doctrine (against birth control) the Irish achieved the necessary outcome by delaying the age of marriage. It is not difficult to cite other examples of social learning.

When expectations drive decisions, and the expectations are fulfilled, it can be presumed that the agent prefers the resulting state of affairs to the initial one, at least instantaneously. That being so, the agent would not choose to return to the initial state. This is tantamount to uni-directionality (irreversibility) of economic decision processes. In or near a Walrasian equilibrium this preference relationship is almost equivalent to the preference ordering assumed in standard neoclassical utility theory. As pointed out previously, only the further assumption of unchanging preferences and path-independence satisfies the integrability condition for the neoclassical marginal utility gradient function, defined on commodity space. Integrability, in turn, is a necessary condition for agents to optimize their choices.

However, the existence of implicit *instantaneous* preferences between neighboring states-of-being can also be applied in a non-equilibrium environment, although the possibility for optimal choice cannot be asserted in that case. Nevertheless, to introduce a scalar function that can be interpreted as individual well-being we need a complete ordering. The avoid avoidable loss (AAL) rule constitutes a local order criterion. The problem is that the economic state-space of the agents is characterized by multiple local-ordering criteria. This is because of the fact that there are several kinds of irreversibilities in economics. The existence of one (and only one) irreversibility would imply the existence of an unique economic equilibrium toward which the system would naturally tend. The existence of others clouds the issue and casts doubt on the existence of such a unique economic equilibrium. Indeed, one of the irreversibilities – the accumulation of knowledge – may tend to drive the system away from such an equilibrium.

The next section discusses the various types of irreversibilities in economics (Martinás 2000). However, it is not essential for the derivations that follow, and may be skipped by a reader in a hurry.

3.2   IRREVERSIBILITY IN ECONOMICS

In the model economy described hereafter several distinct kinds of irreversibility play a role. Each of them requires different theoretical tools. We focus mainly hereafter on the first and fourth types, but note that deeper analysis of either of the first two, or the last, may lead to productive new insights.

3.2.1   Irreversibility of Wealth Accumulation

As mentioned already more than once, competent, minimally rational economic agents conduct sequences of economic unit processes that are (almost) always uni-directional, in the specific sense of increasing their wealth. This occurs as a consequence of the AAL rule, discussed in detail hereafter. Every agent, confronted with a decision possibility, evaluates its wealth in terms of a scalar function of stocks of goods and money, Z. The AAL rule implies that the wealth function must almost never decrease as a consequence of voluntary actions. The agent will never choose to lose. This rule functions independently of physical irreversibility (discussed next) although physical irreversibility must be taken into account. Similarly the AAL rule is independent of the irreversibility arising from learning and experience, mentioned above, even though experience and learning are essential to the avoidance of losses that are avoidable in principle.

This kind of irreversibility was implicitly defined more than a century ago, albeit implicitly, by Carl Menger (Menger 1981), founder of the Austrian School. Menger stated the necessary conditions for free exchange of goods among economizing individuals (belonging to *H. Economicus*) as follows:

a. One economizing individual must have command of goods which have a smaller value to him than other quantities of goods at the disposal of another economizing individual who evaluates the goods in reverse fashion,
b. The two economizing individuals must have recognized this relationship, and
c. They must have the power actually to perform the exchange of goods.

Menger added that the absence of even one of these three conditions means that an essential prerequisite for this exchange is missing, and that therefore the exchange will not take place at all. The idea, in brief, is that a (minimally rational) economic agent never initiates or agrees to an exchange, if that exchange would be expected to result in a loss. Menger's three conditions for an exchange are essentially equivalent to our AAL rule, which is defined more precisely and extended to include production activities, below.

While the AAL rule is clearly implicit in Menger's three conditions, the Austrian School did not develop the mathematical consequences of the rule.[1] However, we note that the form of the AAL rule itself, as applied to any given agent or firm, evolves over time partly as the result of learning from experience in the marketplace and partly in response to changes in the larger system within which the agent operates. In more abstract terms, avoidance of (avoidable) losses from economic transactions or production decisions *implies a strict ordering of allowed states*. This instantaneous ordering follows from the ability of the agent to attach internal monetary values to each state (where a state is defined in terms of stocks of goods and money). This ordering can be expressed as a set of inequalities. The inequality constraint has important mathematical implications, as we show subsequently.

It should be noted that the irreversible increase of individual wealth does not guarantee that social or community welfare will increase irreversibly. On the contrary, as Chipman and Moore showed in 1976, more GNP need not correspond to greater welfare (Chipman and Moore 1976).

3.2.2 Physical Irreversibility

With few exceptions, every economic process is accompanied by energy and/ or material flows. Hence most economic processes are also thermodynamic processes. In particular, certain processes involve waste flows to the environment resulting from inefficient purification, conversion or forming processes, and require compensating extractive flows and environmental degradation. Durable goods suffer wear and tear (thanks to the Second Law of thermodynamics), which ultimately result in the need for repair or replacement. Some other economic processes result in consumption wastes and some processes involve extractive flows to replace these losses, as well as increasing the total stocks of material goods in the anthroposphere. Nicolas Georgescu-Roegen gave a very good summary on the relevance of the Second Law (the so-called entropy law) to economics (Georgescu-Roegen 1971, 1977, 1979; Ayres and Nair 1984). Other useful references include Daly (1985, 1992b); Perrings (1997); Cleveland and Ruth (1997); Ayres (1998).

The laws of thermodynamics imply the existence of an equilibrium state. A closed and isolated system will tend to evolve toward this equilibrium. However an open, non-linear system, in contact with a flow of free energy from outside, can 'self-organize' and remain in or near stationary 'dissipative' states that are far from thermodynamic equilibrium (Glansdorff and Prigogine 1971; Nicolis and Prigogine 1977). The biosphere is an example of such a self-ordered system, far from equilibrium.

We suggest, hereafter, that the flow of knowledge into an economic system is analogous to the flow of free energy in a dissipative system, such as the ones discussed by Nicolis and Prigogine. It may also tend to keep the economic system far away from Walrasian equilibrium, while creating stationary states (local minima) around which the system tends to move. However, the analogy should not be pressed too hard, because knowledge is cumulative, like entropy, whereas the free energy flow that drives biosphere evolution is not. This suggests that in economic evolution there might be a progression of such local minima, corresponding to (temporary) stationary states.

3.2.3 Irreversibility arising from Accumulated Knowledge or Experience

Learning from experience or learning-by-doing requires time and must be conceived as a uni-directional process of accumulation.[2] There are strictly personal (non-economic) kinds of learning by experience, such as how to live with a marriage partner or how to raise children, which we need not consider further. In the economic context, learning or experience improves skills of all kinds, ranging from literacy and numeracy to manual skills such as driving a car or operating a word processor. These 'labor' skills are essential to economic activities, of course.

It is very important to recognize that learning from experience or learning-by-doing can increase economic efficiency only up to a point that is technologically or physiologically determined. For instance, by dint of practice, a very skilled typist can type 120 words per minute (wpm), as compared to the minimum speed regarded as acceptable for a secretary (around 60 wpm) and the speed achievable by typical two-finger typists (20–30 wpm). A skilled bricklayer can lay bricks several times as fast as a journeyman. Examples of this sort can be found in all domains.[3] Manual skills can be taught only to a limited degree and improvement inevitably requires practice. ('Practice makes perfect'.)

However for our purposes, it is important to note that, just as every individual needs to learn how to live in society, an economic agent must also learn how to survive in an imperfectly competitive market.[4] If economic agents are viewed as players in a game, they must learn the rules of the game, the strategies of the game and the best responses to possible moves by other players. Games-playing skills are obviously learned, mostly by practice. The notion of economic activity as gaming has become central since the idea was put forward by von Neumann and Morgenstern nearly 60 years ago (von Neumann and Morgenstern 1944, Nash 1951, 1953; Harsanyi 1962, 1966; Shubik 1982; Holland 1986, 1988; and Gintis 2000b). An extension of the gaming idea, in the real economic context – and in social and political life – has to do, in part, with the evolution of cooperation, as a legitimate strategy option, from pure competition (Axelrod 1984, 1987; Axelrod and Dion 1988). Finding the optimal balance between the two in each particular situation is also a learning process.

Evolutionary economic growth models focus specifically on firm's improving efficiency (technical progress) by searching through a set of existing technical possibilities, subject to given decision rules (Nelson and Winter 1974; Schwefel 1988). However, we also emphasize that technologies are never characterized simply by capital and labor requirements, but depend on materials and materials properties, and flows of available energy (exergy). We also distinguish between a search among known techniques or possible variants of known products (which may lead to adoption and diffusion or imitation) and a search for new knowledge of a more fundamental sort.

An important distinction is between the 'level' (or *quality*) of knowledge and the *quantity* of knowledge in terms of the number of people who share it or books that include it. Knowledge that is abstract and not embodied in people or machines can be shared with others without any loss to the original 'owner' (albeit not without cost). This sharing or diffusion increases the total quantity of knowledge – or 'human capital' – in the system, as noted previously. This makes it fundamentally different from physical commodities. On the other hand, knowledge gained by experience of how to communicate across cultural or social barriers, or how markets work or 'how business works' or 'how to manage people' is not easily shared or diffused, even though business schools try to teach some of these kinds of knowledge.

Certain kinds of qualitative knowledge that is shared by few people may have a high 'exclusivity' value such that the more people who share it the less it is worth to any of them. An example might be knowledge of the location of a gold mine or advance knowledge of the outcome of a battle[5] or a currency revaluation (or devaluation) or an unexpected corporate profit or loss. However, it is important to note that knowledge of the latter sort is not obtained by cumulative learning or experience. The role of experience would only be to enable the agent (most likely a trader) to assess the options intelligently and minimize the probability of loss.

3.2.4   New Knowledge Creation Irreversibility

To a good approximation, the creation of new knowledge is an irreversible process insofar as knowledge once created is rarely lost. (When knowledge is truly lost, as when Cretan civilization was destroyed by a volcanic explosion accompanied by a tidal wave or when the Alexandrian Library was burned or when China was overrun by the Mongols, it is usually due to a catastrophe beyond the realm of economics.) It has been acknowledged for a long time that the primary mechanism for economic growth is innovation, broadly defined to include social and organizational dimensions as well as technology in the engineering sense (Schumpeter 1934, 1912; Winter 1984; Day 1984; Eliasson 1988; Kleinknecht 1987; Andersen 1996).

In the 1950s it was discovered by empirical study that economic growth cannot be explained in terms of increasing stocks of capital and increasing labor supply (Fabricant 1954; Abramovitz 1956; Solow 1956, 1957). The missing factor, really an unexplained residual, was labeled 'technical progress', which is probably another term for increasing productive knowledge. However up to now there has been no way to measure technical progress directly, although some of the inputs (notably literacy, numeracy and R&D) can be quantified.

As noted above, learning-by-experience or learning-by-doing are essential for long-term survival in a competitive environment, and successful operation of the AAL rule. Learning and experience by organizations and institutions can also explain some long-term gains in economic efficiency. That kind of learning yields one kind of economic knowledge. But experience in the sense of cumulative production or multiplicity of interactions does not create new knowledge about nature, nor new inventions, new products or new processes. New knowledge or new ideas may sometimes occur by lucky accident – like the legendary insight into the nature of gravity that Newton supposedly acquired by being hit on the head by a falling apple. But almost invariably they arise from a deliberate search process, whether informal or formal.[6] We discuss the search process later.

3.2.5 Path-dependence Irreversibility

A phenomenon, recognized in evolutionary economics, is known as 'path dependence' or sometimes called 'lock-in'. It can arise when a new technology exhibits very high returns to adoption (David 1975, 1988b, 1988a; Arthur 1983; Arthur et al. 1987b, 1987a, 1983; Arthur 1988a, 1988b). In simpler language, when a technology becomes more valuable and more useful the more people are using it, there is a very strong tendency for the early market leader to achieve a 'lock-in' by making it too difficult for any alternative to get a foothold.

The phenomenon is especially evident when the early leader establishes a standard, for any reason or even by accident, and becomes difficult or impossible to change later because so much capital equipment and so many institutions are based on it. A well-known example is the QWERTY typewriter keyboard (David 1985). The dominance of Microsoft's MSDOS and later 'Windows' operating systems for PCs is a modern example that has aroused passionate debate. A more straightforward illustration is the adoption of the metric system of measurement in Europe vs. the English system in the UK and the US. A less obvious case in point is the decimal (base-ten) system of counting, probably attributable to the fact that humans possess ten fingers. Few people other than mathematicians realize that there are other possibilities that might even be superior in theory, including binary (used in computers), octal (base 8), duodecimal (base 12) or even base 60.

While there are cases where the choice of standard is arbitrary (e.g. driving on the right side of the road vs. the left side) there are other examples where one choice may be significantly better than the others, as for instance the choice of 120 volt AC vs. 220 volt DC current for electric power supply, or the three incompatible video standards (NTSC, PAL or SECAM). While it may be difficult to determine *a priori* which of several possible standards is optimum, the existence of a number of different standards at the same time is clearly undesirable.

Path-dependent 'lock-ins' can, in principle, be broken. The US and the UK are attempting to adopt the metric system, over a period of decades. Europe has adopted a common currency, successfully. Sweden made a policy decision to switch auto traffic from one side of the road to the other some years ago (and achieved it in a single day). However some of the other examples above appear to be unchangeable, because the cost of the change would far exceed the discounted present value of the benefits of the change. Hence, many of the choices that have been incorporated in the material world are effectively irreversible.

The important point is that the 'lock-in' phenomenon is inconsistent with both the neoclassical notion of the technology 'frontier' as a moving boundary in a two-dimensional capital-labor space, as well as the evolutionary model of technological choice as a simple selection among well-understood alternatives (e.g. Nelson and Winter 1982).

3.3   ECONOMIC AGENTS OR ACTORS

It is noteworthy that all of the above irreversibilities, except the first, apply to *H. Sapiens*, as well as *H. Economicus*. (True, *H. Economicus* is a subspecies of *H. Sapiens*, but *H. Sapiens* may operate under rules that do not apply to *H. Economicus*) Humans have attempted to increase their wealth and well-being since prehistoric times. Our species has always had to cope with aging, wear and deterioration (Second Law). People have learned from experience how to hunt effectively, how to defend against predators, how to stay warm in winter, how to conserve food for a period of scarcity and how to secure safety from aggressors and security in old age. Our ancestors also invented and made tools and weapons, such as knives and spears, baked clay pots, bricks, canoes, the wheel, and the sail. Finally, they developed and preserved the knowledge of how to do these things. They even experienced 'lock-ins' (mainly cultural), which preserved obsolete practices, such as circumcision, head coverings for women, food prohibitions, and obsolete organizations (such as feudalism, royal courts, or the Sicilian Mafia) long after their original usefulness had ended.

It can be argued that *H. Sapiens* had to learn, mostly by experience, how to become (at least part of the time) *H. Economicus*. As noted already in Chapter 1, the predominant moral system in prehistoric times, and much of the time since then, has been based on the law of capture, or – to oversimplify slightly – 'might makes right'. The idea that peaceful trade can make both parties better off in the long run than theft, piracy or war had to be learned. The learning process has been long, convoluted, and fascinating, but it is not the subject of this book.

As we have already said, *H. Sapiens* is by no means pure *H. Economicus*. In particular, we humans are not absolutely selfish short-term utility maximizers. We are not even consistent about following the AAL rule – as exemplified by the popularity of lotteries and slot machines – although it is a far better description of what we do most of the time. Yet, some of us still belong to a different species – we have termed it *H. Custodius* – whose members perform other services such as providing physical security, spiritual support, public health and the rule of law, without which our economic system would not last long.

Having said this, it is intellectually convenient for what follows to discover the properties of a system in which the actors are part-time members of *H.*

Economicus insofar as they engage in production and trade of material goods and services associated with material goods. It is these activities that generate the wealth that also supports – via taxes – those of us belonging to *H. Custodius* or just *H. Sapiens.*

In the real economy most production (except of labor services) is carried out by firms, which are organizations consisting of people, capital stocks, stocks of intermediate goods, and money. But firms are more than simple groupings; they have history, reputation, internal culture, organizational structure, policies (including incentives), productive knowledge (in the sense of 'know-how'), and market knowledge. A very small firm may consist of a key individual and some helpers who are interchangeable, but a large firm may have many 'key' employees, and operating systems or protocols, that are not readily replaced.

These features greatly complicate mergers, acquisitions, spinoffs, and hiring and/or promotion policies. The details are beyond our immediate concerns except for one point. It is vitally important to recognize that firms are differentiated from one another in terms of what is now known in business schools as *core competencies,* just as individuals are differentiated by inborn talents and learned skills. In due course it will be convenient to assume that differentiation among firms is partly attributable to their differing stocks of 'knowledge', in a very general sense. Moreover, this knowledge can be accumulated (or, in some circumstances, lost) but it is not bought or sold in small increments in any market.

A final point about firms is that they can be assumed to be profit-seeking, which is another way of saying that they attempt to avoid losses, that is, to operate consistently according to the AAL rule. As with individuals, we do not assume that firms always attempt to maximize profits in the short term, for some of the same reasons that individuals do not. However, we assume that firms value goods in terms of money, whence money and goods are interchangeable for purposes of valuation. (Knowledge, on the other hand, is exchangeable, and hence valued in money terms, only when it is embodied in some physical form, such as software sold in a CD, or possibly when it is considered to be part of the value of a firm, taken as a whole, when the firm sells shares in the stock market or when it is acquired by another firm.)

3.4 WEALTH

For purposes of the next several chapters we assume that the realizable wealth of any economic agent consists only of tangible assets, namely stocks of material goods – from which services may be obtained – plus money. The important role of knowledge is temporarily disregarded, for purposes of exposition. Later, knowledge and other immaterial assets like 'goodwill' or 'synergy' or research in progress will be considered, since knowledge and reputation clearly contribute enormously to the value of individuals (measured in terms of earnings potential), while other intangibles may also add considerably to a firm's market value as measured by the stock market.

For simplicity, then, we treat both individuals and firms as *agents* belonging to the species *H. Economicus*. Each agent belonging to *H. Economicus* normally seeks to preserve and increase its tangible wealth. (Artists, scientists and scholars have other motivations, to be sure.) In the real world, both natural (physical) laws and social norms impose constraints. We explore these constraints, and derive the properties of *H. Economicus* in a systemic context. The main result of this derivation is that there exists for each agent, at every moment in time, a scalar measure of tangible wealth, denoted the Z-function hereafter, which is a function of stocks of goods and money.

In effect, we argue that the casual intuition that wealth consists of stocks of goods and money (or money equivalents) is theoretically sufficient to explain some key economic phenomena. In order to evaluate possible transactions or production decisions, an agent needs to assess the impact of any change in stocks on the wealth function. Based on this function, complex interactions with other agents, and with the natural environment, can be described mathematically, leading to dynamic 'equations of motion', for an economic system. This enables us to characterize, in mathematical terms, the 'driving forces', and the necessary conditions, for economic growth. An endogenous growth theorem is ultimately derived.

A brief word is needed here on the role of money. We refer to *money* in this chapter in the usual sense of cash plus credit (or borrowing capacity).[7] We assume the existence of a money economy and a functioning market in which the money can be exchanged for goods. Money is a medium of exchange which facilitates trade. To each individual economic agent, it is a given. Without money, exchange could only take place by barter, which has very high transaction costs (in labor time) to traders, inasmuch as each trader needs to find one or more suitable partners. Obviously a central marketplace helps to reduce the friction, but only to a degree. Evidently money short-cuts the search process, and thus reduces the search time and increases the efficiency with which buyers and sellers can interact. Money is therefore a lubricant, and – like lubricating oil in a mechanical device – it makes the system work better and therefore has a value to the system. In practice, money is 'created' by the economic system as a whole (or its custodians). We discuss the role of money, prices and the money supply in more detail in Appendix A.

The main actors in the economic system are individuals and *firms*, which are legal persons, but with indefinite lifetime. Firms are buyers of labor and goods and services (raw materials or intermediate goods) and sellers of goods or services to other firms or individual consumers.[8] They produce goods or services from labor, working capital (money) and other goods. They also generate wastes. By comparison, individuals 'consume' products and produce only labor and wastes. The behavior of these economic agents can be described, for the most part, by the same formalism, based initially on the AAL rule.

3.5   STOCKS, FLOWS AND TRANSACTIONS

As noted above, wealth is an accumulation – hence a *stock* – of material goods (possessions), plus money. Knowledge, as such, is not wealth. The material goods provide intangible services to their owners. The actors all have preference orderings for the services associated with goods, which translate into preference orderings for the goods and goods-related services. Stocks of goods are tangible. Laws of nature (notably the laws of thermodynamics) constrain their changes. If these stocks are measured in natural, physical units, then the physical constraints will be transparent.

For example the measurement of a stock can be done in kilograms or in energy (actually *exergy*[9]) units. Mass (kilogram) or energy units have the great advantage that the most basic law of nature, arguably, is the law of conservation of mass and energy.[10] Mass-balance (and exergy-balance) laws are strict book-keepers. There are no known processes violating the law. If the apparent balance is not perfect, we assume that there is a measurement error or a mistake in the description of the process. Where possible we exploit the mass/energy conservation rules by writing the balance equation for stocks. The mathematical details are set forth in Appendix B.

For an actor (or economic agent) the voluntary 'unit operations' which result in stock changes include consumption, production, trade and gifts/ donations (incoming or outgoing). Involuntary unit operations include taxes and losses from disaster, theft, and/or wear & tear. In all cases there is a physical conservation law that must be satisfied. We discuss them in specific terms, briefly, as follows:

Trade   refers to exchange of goods or services (including labor) for money, or goods for goods (barter). The quantity of material goods traded can be measured in physical units (i.e. mass or exergy) as well as monetary value. Except for the case of labor for money, the total change of stocks in mass terms is always zero because an exchange transaction changes only the ownership, and possibly (but not always) the physical location of the material stocks. As money and labor have no mass, the mass conservation law implies that the reduction in seller's stocks is balanced by the increase in buyer's stocks, measured in mass terms. The total change in money terms is also always zero. If third parties are involved (for instance, if there is a tax) then the money balance still holds if all changes are counted. Needless to say, transaction costs (including time, taxes, fees, commissions, etc.) are rarely zero in the real world.

Production   in economics normally implies value added to a commodity or service through the application of labor, capital services and other commodities, including exergy in some form. The notion of value (as in 'value-added') seems to imply that value is an attribute of every produced good. This implication is superficially inconsistent with our understanding, elaborated later, that value is determined by the preferences of consumers. The point will be elaborated and the apparent conflict reconciled later. In the context of production theory, we take 'value' to mean *market value*.

Production has two variants, depending on whether we are talking about goods or services. The first variant refers to the transformation of stocks of raw materials or purchased intermediate inputs from the initial form (as purchased) to a higher value form suitable for sale to another producer or final consumer. Production in this context involves inputs of labor and capital services and flows of commodities, including exergy.[11] The second variant applies to the production of services, including the production of labor by individual workers. The production of services is inherently linked to the consumption of those services, since services cannot be stockpiled but must be used immediately.

The quantity of material goods transformed or consumed can be measured in physical units (mass or exergy) as well as monetary value. The stock change of material inputs to production is negative, while the stock change for material products, by-products and wastes is equal and opposite in sign (i.e. positive). There is always some non-zero loss (waste) during every transformation process. Agriculture, fishing, forestry and mining are special cases where the raw materials actually embodied in the products are actually 'free gifts' of nature (although capital and labor are required to utilize them.) We discuss these cases subsequently in more detail.

It is very important for what comes later to recognize that, although we can describe a production process in terms of stocks and stock changes, modern industrial (non-artisanal) production involves very large batches or continuous flows. The image of an artisan slowly carving a horn to make a spear point, or making a pot from wet clay, is now obsolete. Equally irrelevant, to be candid, is the picture present in many modern economics textbooks, of many independent competing producers making the same product (e.g. bread) from (purchased) capital and labor services, but without inputs of flour for the dough or fuel to heat the oven.[12] Nowadays we envision a line of trucks or railway carriages unloading several different raw materials at one end of a plant, a line of workers entering or leaving through the front door, and another line of trucks or railway cars carrying finished products (and waste products) away from the other end of the factory.

The modern factory, like the modern household, depends upon inputs of so-called utility services, such as electric power, telecommunications, water and sewer services, transport, telecommunications, financial services, accounting, legal and insurance services and others that are needed for virtually every economic activity. Moreover, many of these services are *flows* and, in most cases, it is difficult and inconvenient, or impossible, to reconvert the flow back to a stock. Electric power can be stored briefly in storage batteries, and water can be stored in cisterns, but these are exceptional cases. Generally speaking the transformation from discrete stock changes to a continuous flow is a one-way process.

What has already happened to utility services is in the process of happening to most products, except structures. Batch production of discrete units or quantities by workers using individual tools or individual machines is being replaced in many manufacturing sectors by continuous production lines in which the workers scarcely touch the products and merely monitor and maintain the machines. In these cases utility services, plus continuous flows of raw or intermediate processed materials, together with labor and capital equipment, generate a continuous stream of products. These products are distributed to users and consumers by means of transport and wholesale/retail services utilizing labor and capital equipment, as well as utility services.

Still other sectors produce intangible services (like insurance, banking or entertainment) from labor, capital goods and utility services. In short, the production system now consists of many sectors, some of which produce flows of utility services from stocks of raw materials (and machines), while others (such as construction) still produce discrete products from flows of utility services and raw material inputs, while still others produce intangible services from labor, utility services and capital goods. This system cannot be characterized adequately in terms of a simple single-sector model where all agents produce a single homogenous product (e.g. 'bread').

Consumption may be *intermediate* or *final*. *Intermediate* consumption is a stage in the production chain. *Final* consumption is the end-use of a product or service to increase welfare or well-being; it is the objective of organized economic activity. In modern economies the consumption of home-made goods seems to be negligible, although it is by no means unimportant in some developing countries. The producer and the consumer are distinct economic agents and the exchange transaction is almost always monetary. Consumers may be wage-earners who receive money for labor, or they may be pensioners, who live on income transfers. (Young children do not make economic decisions, hence they are not economic agents for the purposes of this book.)

Consumers may accumulate stocks of material goods, but stock changes due to dissipative usage of materials *per se* are always negative. Such uses are pervasive. They include fuels, lubricants, fertilizers, solvents, soap and detergents, acids, abrasives, pesticides, cosmetics and so on. The conservation of mass implies that the corresponding 'stock' of mass in the environment increases by an equal amount. Consumption is also normally an AAL process for *H. Sapiens*. For *H. Economicus* consumption is a loss.

The two basic operations (trade and production in the sense of value added by some combination of labor, capital services and other commodities) are called *economic unit processes*. We must, nevertheless, recognize the possibility of other non-economic processes, including tax payments, loss to theft, fire, flood, or other disaster, gifts and donations (in or out). All of these events also result in money and/or goods stock changes. Moreover, physical processes such as aging, deterioration, dissipation and wear also reduce stocks of goods.

Some of the stock changes noted in the previous paragraph are involuntary, and some are voluntary. The involuntary changes, such as fires or floods, can be large at some times and places. The effect of involuntary changes is essentially always negative. If involuntary processes were the only ones to consider there would be no economic life. (Voluntary gifts and donations of money are sometimes large for individuals, but these seldom involve changes in material stocks, and we can safely neglect this possibility, at least for the present.)

Neglecting the possibility of free gifts from other economic agents (but not so-called 'free gifts from nature'), there are only two ways in which the stocks of material goods in the possession of an agent can increase. One of them is by acquisition by purchase (exchange of money for goods) and the other is by production from intermediate goods already in stock or from resources extracted from the environment by means of labor and capital. In the case of acquisition by purchase, the money stock of the purchaser decreases. In the case of production from other goods, the stock of raw or intermediate materials decreases. In the case of extractive resources the stock of underground resources (e.g. hydrocarbons, metal ores) decreases. Only in the case of production from common property resources (air, water) are the producer's stocks (of land, fishing boats, tools and equipment) left unchanged except for erosion and physical wear.

3.6 Q-CYCLES AND VALUE-ADDED

Observing the stocks of a normal agent over time, it appears that stocks of goods and money are always changing. Some are increasing, some are decreasing. Unless all stocks are continuously given equivalent monetary values (based on a market price) it seems to be very difficult to decide whether the agent is gaining or losing wealth from moment to moment. Nevertheless, that judgment is possible by making use of *quasi-cycles*, defined below.

A '*quasi-cycle*' or *q-cycle* is a sequence of economic unit operations resulting in a state in which the original stocks of materials are reconstituted as they were originally; the money stock is either larger or smaller than before. We denote the first case as a *profit quasi-cycle*. The second case is a *loss quasi-cycle*. For convenience hereafter, we simply refer to a q-cycle, since the sequence rarely recreates the original set of stocks and money exactly.

To illustrate, imagine a factory that produces widgets from two types of purchased components, namely 'bits' and 'pieces', plus labor and utility services. It sells the widgets to a distributor. From the money received in payment, the firm pays for labor and other utility services, and buys more components to replace what has been used up. Suppose also that the firm gets an offer to sell 500 widgets for $10 000. Labor and other fixed costs account for $4000. The firm receives an offer from a supplier to provide enough 'bit' components for 500 more widgets, at a total price of $3000. The manager accepts this offer. This transaction leaves $3000 available to buy 'pieces'. If the firm can buy enough pieces for 500 widgets for $3000 dollars, the q-cycle is closed – no loss, no gain. If the price of pieces has risen so that the firm must pay $4000 to buy enough for 500 widgets, the firm will lose $1000 in that cycle. If the price of pieces falls, so the firm can replace its piece supply for $2500, there is a gain (profit) of $500.

Obviously there are many variants of this scenario. The prices of inputs (bits and pieces) can rise or fall. The cost of labor can increase (e.g. there may be a strike). The market price of finished widgets may rise or fall. The firm's managers necessarily operate on the basis of short-term expectations. There are unavoidable contingencies. The cleverest manager in the world cannot avoid them all.

For the sake of generality, we can extend the term q-cycle to include cases where some of the original stocks are recreated exactly while some others are either increased, or decreased. The agent is richer if all the stocks are the same or larger than in the initial point, and poorer if all the stocks are the same or less than before. The mixed case, where some stocks have increased while others have decreased, is really an incomplete cycle: it can be completed (at least in principle) by buying more of stocks that are deficient and selling the excess stocks of other goods, until there are only deficits or excesses compared to the original state, but not both.

We acknowledge that this hypothetical procedure does not necessarily lead to a unique result since it depends upon the existence of market prices, which (as noted already) can change from moment to moment. As it happens, the wealth function Z (formally defined later) resolves this difficulty by providing a measure of wealth that is unique to each agent, provided the agents' preferences remain unchanged for as long as the cycle takes to close.

For a firm, the standard q-cycle involves the purchase of labor and utility services, purchase and conversion of raw materials to finished products, and sale of the finished product or services to recover the money spent, replace the materials used, and, of course to pay interest and taxes. In principle, the materials in inventory at the end will be the same as at the beginning, but the amount of money in the bank will be larger if the firm is operating at a profit, and conversely, it will be smaller if the firm is operating at a loss. Labor and utility services cannot be stockpiled, of course, as already noted.

For an individual, the standard q-cycle involves purchase of consumables (for money), use and dissipation or discarding of the consumables, sale of labor, and receipt of money payment for the labor (wages), thus replacing the original money stock. If the wages received exceed the individual's requirements for subsistence plus involuntary losses, such as taxes, the individual is able to save some money. This can be used to purchase other goods and increase stocks, or to save and invest. Either way, the individual usually becomes wealthier. If there is no surplus, however, the individual becomes poorer.

A *process chain* is a sequence of linked q-cycles, beginning with the extraction of raw materials and ending with a finished product. If each step in the sequence satisfies the AAL rule, the sequence as a whole will yield a product or products worth more than the sum of all purchased inputs (including purchased labor) along the chain. The surplus value can be regarded as profit, rents (payments to the owners of natural capital) and payments to produced capital services. The sum of all payments to labor and payments to capital can be equated to value-added. The notion of value is discussed in more detail later in this chapter and the next. For a product or commodity that is sold in the marketplace, value and price can be equated.

It is evident that economic growth is possible if, and only if, most q-cycles result in positive gains. However, everyday experience indicates that negative (loss) q-cycles actually do occur in real life, and not all that infrequently. Some of these losses – losses of the first kind – are unavoidable; others, of the second kind, are avoidable. The first kind results from the fact that actions take time and are subject to inherent uncertainty and the difficulty or impossibility of realistic planning. This kind of loss is exemplified by a futures trader (or arbitrageur) in some commodity, such as wheat. The trader pays cash in advance to the producer for future delivery. He relieves the producer of the risk of market fluctuations and expects to be compensated for this by making a profit on the future sale of the commodity at a higher price. He works on the basis of averages. His profits fluctuate with the wheat market. But occasionally something happens in the market between the time he buys the wheat futures, and the time he takes delivery. If the market price has fallen sharply (e.g. because of an unexpectedly large harvest in the Ukraine), the trader will lose money.

Another example of unavoidable losses is worth mentioning, since it is critical for understanding technological innovation and economic growth. Technological change on the macro-scale is the result of a continuous stream of innovations by entrepreneurs. But successful innovations often – perhaps more often than not – harm established businesses. Schumpeter characterized this phenomenon as 'creative destruction'(Schumpeter 1912, 1934) – another word for unavoidable q-cycle losses.

The statistics of venture capital firms confirm that the majority of start-ups fall by the wayside within a few years. The great successes are extremely rare. Yet it is impossible to know at the outset which ventures will succeed and which will not. Hence, it is not irrational for inventors to try to make a fortune from their inventions, or for venture capitalists to finance them. Failures and losses are normal and expected. They can occur for many innocent reasons, ranging from accidental glitches in a prototype to under-estimates or over-estimates of the market potential, competition from a better capitalized rival, or just being a little too late with a sound idea. It is only required that the occasional great successes pay for all the losses.

It is a fact that most new businesses fail, whether the new business is a shop, a restaurant or a high-tech startup. The failures of high-tech startups are mostly unavoidable, for reasons mentioned above. It can be argued that losses of risk capital are an essential feature of innovation and hence necessary for a growing economy. However, most of the small shops and restaurants fail for other reasons, for example, because the entrepreneur is under-capitalized, in the wrong location, lacks essential information about suppliers or customers or the product itself, or lacks essential management skills. These losses are mostly avoidable in the sense that a more experienced entrepreneur would have done it differently or not at all. In short, the necessary skills can be learned.

In order to survive (in the economic sphere), an economic agent must learn to avoid such avoidable miscalculations. A restauranteur must be able not only to cook well, but also to buy good quality supplies, attract customers, and keep books. A shopkeeper must not order too much or too little, or goods that are too costly or too low in quality. An entrepreneur must not borrow more than can be repaid out of normal cash flow with a safety margin. And so forth. We can assume that the learning process requires a small – certainly finite – number of cycles and that, thereafter, the economic agent is able to avoid any further losses. This degree of skill and experience may be termed 'minimum rationality' since it is even less demanding than Simon's 'bounded rationality' (which allows only for imperfect information). It is far less demanding than 'perfect rationality' in the traditional neoclassical sense.

In summary, the first kind of (unavoidable) loss arises for reasons beyond the control of the agent, such as extraordinarily bad weather, or sudden competition from a more powerful competitor, or when there is no organized market – as opposed to latent demand – for the goods (or labor skills) being offered for sale. Losses of this kind are a consequence of human inability to predict the future beyond a short-term horizon. Losses of the first kind are essentially unavoidable, but they are rare enough to allow the macro-economy to function. In fact, it can be argued that if such losses are an inherent consequence of risk-taking, and that if entrepreneurs were unwilling to take any risks, there would be no investment, and certainly no innovation.

The second kind of (avoidable) loss arises when the agent makes bad decisions that a more experienced agent would not make. Such losses can be, and must be, avoided by a normally competent economic agent, which is what we mean by obeying the AAL rule. Otherwise the agent's stocks of goods and/or money will decrease below a critical level, and the agent (if a firm) will go out of business or (if a worker), will become unemployed. In short, the actor will cease to be an economic agent in the sense of our model.

To recapitulate earlier comments, we argue that economic agents, whether firms or individuals belonging to *H. Economicus*, try to avoid avoidable losses that decrease their wealth. More precisely, we now assert *that a competent, minimally rational economic agent is one who consistently avoids q-cycles of the second kind.* Moreover, we assert that *avoidance of loss (q-cycles of the second kind) implies the essential irreversibility of economic activity.* The idea is that agents that consistently fail to break even or better disappear quickly from competitive entrepreneurial economic life (except perhaps as workers or agents employed by others), whereas consistent winners increase their own wealth and that of society as a whole.[13]

For the economic system as a whole, it is evident that a necessary condition for sustainability in the economic sense is that value-added (in the sense of market value) minus payments to labor should be sufficient to provide a surplus to compensate for all capital depreciation, including both produced capital and natural capital. Obviously economic growth requires that the surplus value added be sufficient to enable additional investment.

The obvious analogy with Darwinian evolution through 'survival of the fittest' may be imperfect, but it is nevertheless useful. In fact, we argue that minimal fitness in the above sense, rather than maximal fitness (maximization), is the criterion for economic survival. The evolutionary perspective adopted here is essentially identical to that laid out previously by evolutionary theorists, such as Richard Nelson and Sidney Winter. For instance, consider the following:

> The key ideas of evolutionary theory have been laid out. Firms at any time are viewed as possessing various capabilities, procedures, and decision rules that determine what they do given external conditions. They also engage in various 'search' operations whereby they discover, consider, and evaluate possible changes in their ways of doing things. Firms whose decision rules are profitable, given the market environment, expand; those firms that are unprofitable contract. ... (Nelson 1982, p. 207)

We share that view.

3.7   UTILITY MAXIMIZATION VS. THE AAL RULE

An exchange in which both agents maximize their utility evidently satisfies the three Menger criteria, quoted earlier. This, in turn, guarantees the irreversibility of economic activity and the eventual approach (after many exchanges) to a Walrasian equilibrium. However, we have observed that the AAL rule is considerably weaker (i.e. less restrictive) than the utility maximum principle. It is a truism, that if Jill is ready to trade an apple for an orange, then she will not be willing to trade an orange for an apple (unless it is the activity of trading *per se* that she enjoys). To put it in the formal language of utility theory, her preference for consuming an orange must be greater than her preference for consuming an apple. Thus preference relations are not only transitive but, by the same token, they imply irreversibility.

Still the traditional concept of utility has several disadvantages. First of all, it is directly applicable only to consumption. Yet not only the consumer but all economic agents, including firms, have to be able to choose among the actions available to them, and choose those actions that improve their economic state and avoid actions that impair it. This ability can be characterized as the AAL rule. It presupposes that every economic agent has the species of rationality that we call 'common sense', and that it is capable of learning from experience and modifying its expectations. It does not require perfect rationality, or perfect foresight and it does not necessarily follow that the actions taken are optimal in the usual sense.

A further disadvantage of canonical utility theory is that it is inherently *myopic* in the time dimension. It is easy to prove that a static maximum is very different from a dynamic maximum, if such a thing can be satisfactorily defined at all. Maximizing the utility of current consumption is not necessarily consistent with maximizing long-term welfare. Economic theorists have evaded (but not avoided) this problem by converting a dynamic maximization into a static maximization, by maximizing an integral over time. The trick is to introduce the notion of maximizing the *present value* of future consumption, using a discount rate, which is usually taken to be a rate of interest or a rate of GDP (or, more usually, consumption) growth. Of course, the choice of discount rate for any given calculation implies the ability to foresee the future. In reality, the choice is often almost arbitrary.[14]

In real economic life, most agents – whether individuals or firms – cannot perform this neat trick. What they can do to ensure future as well as present consumption is to save and invest. Sending children to school instead of to work is a way of investing in their future, and possibly also providing for old age. Buying a house instead of paying rent is a device for saving, as well as providing incentives for property improvement, especially via 'do-it-yourself' or DIY. Nowadays, there are a number of more explicit mechanisms for trading some present consumption for some future consumption. However none of these schemes can guarantee a specified future standard of living, still less maximize that future standard of living.

Nevertheless, it is evident that for every agent (individual or firm) there is a minimum finite bundle of goods and money necessary to ensure economic survival, barring a catastrophic unavoidable loss of the first kind. Since this is another way of saying that zero wealth is not sufficient to assure survival, it is a fairly trivial assertion. However this assertion is important for the chain of logic we are constructing.

In summary we argue that, in practice, economic agents (i.e. members of the species *H. Economicus*) try to maintain and increase their wealth, not (only) their current consumption.[15] In other words, current consumption is a function of wealth, not *vice versa*. There are some exceptions, mainly in the extreme case where current consumption is below the subsistence minimum, but the exceptions do not disprove the general rule. Humans accumulate stocks of goods for future use, just as squirrels accumulate stocks of nuts for the winter or Canada geese accumulate a layer of fat to fuel their long migration. The tendency of animals to accumulate is genetically determined. However a crucial difference between biology and economics is that we humans are able to make conscious tradeoffs between consumption and accumulation, based on intelligent evaluation, even though true optimization is far beyond our capabilities.

3.8   ON THE IRREVERSIBILITY OF UNIT OPERATIONS AND THE UNI-DIRECTIONALITY OF QUASI-CYCLES

As mentioned already, unit operations and quasi-cycles take finite time to complete. Some operations are fast enough to be considered instantaneous, but not all. In fact some operations can be quite time consuming (consider the processes of wine and whiskey-aging, for instance). What matters is that changes can occur in the environment (including the 'state of the market') during the time lapse between the beginning and end of a q-cycle.

Another important point to remember is that each decision by the agent is not for a q-cycle as a whole, but for the elements of the q-cycle. These elements are defined as unit operations. Yet the agent must decide on, and perform, each action in the sequence subject to the overall requirement that the final (quasi) cycle results in a gain, not a loss. This requirement is equivalent to irreversibility. The decision to proceed from one state to another precludes a decision to proceed in the opposite direction.

The decision has another strong implication: It means that the agent (individual or firm) must have, and must act on, *expectations* about the (immediate and short-term) future of its immediate environment, including the behavior of trading partners and competitors. The AAL-cycle rule therefore implies that agents have coherent expectations, based on experience and knowledge. Moreover, it implies the existence of a scalar function (of stocks of goods and money) which we call the wealth function.

The argument proceeds as follows. Only q-cycles resulting in gains are allowable, by assumption. In a gain cycle every unit operation is expected (by the agent) to be feasible. In a loss cycle there must be at least one unit operation that is non-feasible, hence must be rejected, based on the agent's set of expectations. During the learning process, when expectations are being formed, a loss experience modifies the agent's expectations, so as to avoid future losses.

Among all possible q-cycles applicable to a given agent, there is a set of *gain* q-cycles and a set of *loss* q-cycles. By definition, these two sets are disjoint. But there is a theoretical possibility of a cycle such that, after some sequence of economic unit operations, the agent reconstructs the original set of stocks exactly. This constitutes the boundary between the feasible (gain q-cycles) and non-feasible (loss q-cycles). The complete cycle is reversible, by definition, whence all the unit exchanges constituting the complete cycle must be reversible. All other quasi-cycles, and their constituent transactions, are irreversible in the directional sense we have defined above.

3.9   PHYSICAL LAWS AND THE AAL RULE

In the real world, as we have pointed out, natural processes of aging, wear and (especially) dissipative intermediate uses, tend to decrease the stocks of all material goods in the possession of agents. Thus the theoretical complete cycle described in the previous paragraph is possible only if there are enough economic gains to compensate exactly for these natural losses.

On the other hand the possible actions of agents are also limited by physical laws. The ability of an individual to sell his/her labor depends upon being healthy, which depends upon physiological capabilities such as the need for sleep and the availability of food, drink and shelter (among other requirements.) Note also that unsold labor is irreversibly lost. It cannot be stockpiled like physical goods.

Similarly, production possibilities are limited by natural laws, including the physical availability of raw materials from natural sources, the properties of those materials and the state of technology for transforming crude materials (including fuels) into finished materials and physical work. Transformation technologies must satisfy the laws of thermodynamics.

Trade is also subject to physical limitations. One of the most obvious (but often forgotten) is Menger's first condition, namely that the agents must have physical command over the goods in question, and his third condition, that the agents have the power to exchange them. It is easy to assume that these conditions refer to legal or institutional arrangements, which they do. But physical conditions are also needed. Trade implies both ownership and control. The latter implies the ability to exclude access by others. To put it crudely, this condition implies the existence of permanent records and markers for land, guardians or fences and/or brands (for animals), physical structures with locks or guards for storage of other goods, and transport facilities to move portable goods from one place to another.

For one reason or another, these conditions do not apply to certain common property natural resources, such as the air, airborne particles such as pollen or dust, the oceans, flowing water, wild animals such as birds, flying insects, fish in the open seas and scenery or climate. In some cases it is possible to circumvent the impossibility of physical possession – hence ownership – of the good itself by enforcing limited access to the good. Beach property, water rights and hunting or fishing rights are examples. But each of these mechanisms depends on the existence of effective private or social enforcement mechanisms with a technological content.

Information (in the sense of knowledge) is peculiar. In the first place, it is immaterial. It is possible for an agent to gain information by means of specific activities, including learning/education and research. Information undoubtedly has value. But it is more often than not a public good, precisely because it cannot be contained or controlled and sold, except in special cases. Software packages, newspapers and books are misleading examples, because the seller does not lose the knowledge and the transfer is not revocable. For this reason, agents with valuable information tend to try to create and protect a monopoly by various means, from patents and copyrights to secret codes and secret societies.[16]

Information (plus labor and, usually tools of some sort) can enable an agent to add market value to material goods. This is one of the most important mechanisms for achieving economic gains. It is only necessary that the (market) value-added exceeds the natural losses. Moreover, there is no physical conservation law for pure information, which means it can be sold more than once – indeed, any number of times. On the other hand, information embodied in material form (such as a book, a computer chip, or a machine design) can be protected more easily. Needless to say, the theft and protection of information have both become significant economic activities in themselves.

A further complication is that many services can be defined in terms of information dissemination. This is obviously true of entertainment, but on reflection it is also and equally true of any service that can be expressed in sensory form, from temperature control to taste and odor (of food and drink), musical experience, visual experience (recreational travel, architecture) and – not least – sexual experience. No theory of economics can ignore information transfer, nor the role of useful information (knowledge).

Technical progress has altered the possibilities for ownership and possession of goods, and of information. Weapons, since spears and arrows, have always played a role in defending property. The same is true of walls. Barbed wire (for fences) which kept herds of cattle or sheep in, and predators out, has been credited by some historians as a major factor in the development of the western US. Satellites have created the possibility of geographical positioning systems and monitoring systems, which might eventually enable governments to control access to areas of the ocean. Printing and subsequent communications technologies encourage the spread of information (as well as disinformation.) Encryption technology is an important contribution to information protection. A social invention, the notion of patenting an invention, was important because it encouraged publication in exchange for a legal monopoly for a limited period of time (Samuelson 1954; Nelson 1959).

There are physical constraints on all economic unit operations. These constraints imply limits on economic activity at all scales, from micro to macro. Moreover, we can say with absolute certainty that economic activity cannot exist without a material basis and an exergy input, to compensate for the natural loss mechanisms mentioned already. It follows that 'gifts of nature' are truly essential to economic activity (and human survival).

This statement is not as strong as it appears, however. The need for some physical materials in economic activity – to be embodied in structures, machines and so on, not to mention food – does *not* impose an absolute minimum (except in the case of food, water and air), because there is no absolute upper limit to the economic services that can be extracted from a physical unit quantity of materials or of available energy (exergy) (Ayres and Kneese 1969).

Dematerialization (including recycling) – in the sense of reducing the material content, as well as the indirect material dissipation associated with production of goods – may continue indefinitely, in principle. The earth is composed of materials, (and there is much more mass floating around in the solar system). Some elements are scarcer than others, but there is no absolute limit to substitution or recycling possibilities. The limiting factor is exergy availability, and the sun provides plenty of that to be going on with. Again, even if the earth's surface were not enough, it is perfectly possible to harvest solar energy from the moon, from satellites, or from space itself.[17] In this regard, we do not agree with the technological pessimism of Georgescu-Roegen (Georgescu-Roegen 1971, 1976; Daly 1992a, 1992b).

In short, economic transactions, other than pure exchange, are not a 'zero-sum game'. This is partly because there are 'free gifts' that can be harvested from the environment (so to speak) by some economic agents, and partly because human ingenuity – expressed as discovery, invention and innovation (discussed above) – enables these gifts of nature to be exploited with ever-increasing effectiveness. Summarizing the last two sections:

1. Consumption of goods is irreversible for both physical and economic reasons. Even if the consumption process could be reversed physically (as when a rental car is returned to the rental agent in 'perfect' condition), it would nevertheless be true that if the agent is willing to make a decision to consume food or beverages or to use some other product that is economically equivalent to consumption, the same agent would not be willing (in principle, of course) to reverse the choice, that is, to go from the state of having acquired and consumed to the state of not having consumed.[18]
2. Production of goods is also non-reversible in some cases (e.g when a metal ore is reduced to pure metal, or when two pieces of metal are welded or a computer chip is impressed with circuitry) and revocable in other cases, for example, assemblies can be disassembled. However production is economically irreversible insofar as if an economic agent is willing to make the decision to produce, he/she would not under the same circumstances make the decision (in principle) of going from the state of having produced to the state of not having produced but having the raw materials and labor ready and available.

3. Trade in goods is irreversible in the economic sense, unless preferences change. (In the short term, at least, it is usually – but not always – reasonable to assume that they do not.) If an agent is willing to buy a good at a specified price, it can be inferred that the agent would not be willing to sell the same good at that price, and conversely.

NOTES

1. Mirowski comments that the work of Menger (and Irving Fisher) was inconsistent with the 'field' formalism that was coming into vogue, and that, in their treatment of production, utility was imputed backward from final goods to factors of production, almost as if it were a value substance embodied in goods (Mirowski 1989b, p. 295).
2. Learning-by-doing is a phrase popularized among economists by Arrow (Arrow 1962), but the term 'experience' as in 'experience curve' is more common among engineers and management consultants. The two terms are essentially interchangeable.
3. However the so-called 'experience curve' at the industry level normally involves other factors, such as economies of scale and process improvements that may also encompass major changes in the technology of production (e.g. Alchian 1963; Wright 1936; Baloff 1966; Rapping 1965; Hartley 1969; Rosen 1972; Argote and Epple 1990; Stokey 1986; Ayres and Martinás 1992).
4. Real markets bear little resemblance to the 'perfect market' of theory. The importance of uncertainty has received attention in the mainstream literature (e.g. Radner 1968; Fehr et al. 1997). The fact that firms are entities composed of real people who do not always behave according to the strictures of Solow's 'trinity' is only one reason. The existence of a variety of institutional constraints is another. Perhaps the most obvious, and possibly the most important, is the fact that the competitive strategy of most firms is to achieve some degree of monopoly, by hook or by crook, precisely in order to limit competition. It is no accident that most industries are oligopolies, and that supposedly competitive firms are restrained from cooperation (i.e. the formation of cartels) only by government regulation and enforcement.
5. The British branch of the Rothschild family made a fortune by exploiting advance knowledge of the outcome of the battle of Waterloo, first by spreading rumors that the battle had been lost (during which period they bought securities cheaply) and subsequently selling when the news of the true outcome arrived a day later.
6. The legend that Fleming discovered penicillin by accident is just that. He was actively searching. Moreover, he failed to follow up his discovery, which was later developed and operationalized by others (Waller 2004, pp. 222–45).
7. Since most borrowing is secured by other assets, it is theoretically possible for the total net worth of an individual or firm to become negative, for example, if those assets are destroyed or lost. However when this happens the individual or firm normally ceases to be an economic agent, for practical purposes.
8. Family firms (at least small ones) usually do not buy labor explicitly from family members, but they share and mutually consume the surplus created by their joint labors, so the effect is the same.
9. Exergy is what most people mean when they speak of energy. It can be defined as 'useful' energy, or 'available' energy. However, energy is conserved in all processes, whereas exergy can be used up. Thus exergy inputs must be balanced by exergy wastes.
10. Neglecting relativistic effects.
11. Georgescu-Roegen has particularly emphasized the fact that production involves both *stocks* (of capital and labor), and *flows* (of materials to be embodied in the product, and useful energy to drive the process (Georgescu-Roegen 1976). *Stocks*, in his conceptualization, are unchanged by the production process, whereas *flows* are consumed.

12.   This is not a joke. In a recent textbook, *Macroeconomics* by N. Gregory Mankiw, a model
      economy of bakers making bread from purchased capital and labor services was invoked
      to explain the theory of allocation of income (Mankiw 1997).
13.   Consumers unable to sell their labor formerly starved. Nowadays they become welfare
      clients.
14.   The US Congress has even tried to legislate discount rates, for purposes of cost-benefit
      (C-B) analysis. In the 1960s and 1970s the rule was that the discount rate for purposes of
      C-B analysis was to be set equal to the lowest rate on long-term government bonds
      outstanding at the time. This rule tended to understate discount rates and exaggerate the
      benefits of long-term capital projects such as dams and highways. In recent years discount
      rates have been set much higher, with the effect of overvaluing the benefits of short-term
      development projects and undervaluing the long-term environmental harms that may
      result.
15.   People such as academic economists, research scientists, teachers and artists may not be
      trying to increase either their physical wealth or their consumption of normal goods. Yet
      they (we) are not members of *H. Custodius* either. Such people are often satisfied to get a
      living wage, or a little more, because their real payment is in the currency of professional
      recognition or pleasure in the job itself. This does not fit very well into economic theory,
      but since the numbers of people involved are small, the exceptions may be considered
      second order.
16.   As noted already, information and uncertainty also play an important role in the operation
      of markets. Asymmetric information is an especially important distorting factor. These
      topics have been discussed extensively in the mainstream literature.
17.   There are ultimate constraints on the amount of thermal energy that can be re-radiated
      from the earth's surface – to keep temperatures from rising too far – but there are also
      engineering approaches to minimizing this problem. It is difficult to pinpoint ultimate
      limits, except to say that they are very far away.
18.   Compulsive eaters, alcoholics, smokers or drug addicts are irrational in this respect.
      Evidently their utility functions are not based on unchanging or very slowly changing
      preferences, but change from hour to hour depending on physiological signals. This is
      another difficult problem for normal economic theory. We (like other economists) neglect
      such cases.

4.   The Z-function

4.1   ON THE EXISTENCE OF A WEALTH FUNCTION

To recapitulate what we have said several times above: economic survival for an economic agent or firm depends upon avoiding losses that can be avoided in principle (losses of the second kind). If, and only if, gain cycles are much more frequent than loss cycles can the firm expect to survive in the long term. Unavoidable losses (of the first kind) will occur from time to time, as we have noted. But survival implies that such losses must be exceptional, and – for analytic simplicity – we can rule them out altogether for the first phase of the following analysis.

The economic survivability conditions for an agent are only testable *ex post*, when the q-cycle is closed (whether the outcome be a gain or a loss). Meanwhile, without foresight of the outcome, the agent has to make a sequence of smaller decisions with regard to unit operations, also based on expectations grounded in past experience and the knowledge stock. For convenience, we call this learned behavior 'obeying the AAL rule'. We will show that the AAL rule implies the existence of a scalar wealth function Z of which the arguments are goods and money. We show that the AAL rule implies that, ruling out occasional losses of the second kind, the economic process is non-decreasing and irreversible, that is, that $dZ > 0$.[1] Allowing for occasional losses, it still follows that market survival means that Z must increase on average.

We have acknowledged that knowledge has economic value, but only in the sense of enhancing the ability of its possessor to make decisions that tend to increase its Z-function. Knowledge of the market will enable the manager of the widget manufacturing firm to select reliable suppliers and distributors and to negotiate prices that permit both parties to make a profit. (A supplier or distributor who sells too cheap or pays too much will soon disappear, with adverse consequences for its trading partners.) It will also help the manager to anticipate market changes, including increases or decreases in demand. Knowledge of the manufacturing process will enable the manager to make intelligent investments in new capital equipment or alternative designs, improved assembly techniques and alternative sources of components. Finally, knowledge facilitates discovery and invention, which improve the long-term prospects of the firm in competition with others. Success in the marketplace therefore depends upon the store of relevant knowledge on the part of the firm's managers and employees. However, in the short term, the knowledge stock embodied in people and organizations can be assumed to be given, fixed and not affected by the outcome of the q-cycle. Knowledge embodied in copy-protected products – including software – can be regarded as another form of commodity. However knowledge in the broader sense is not a commodity that can be exchanged in the market, hence it is not an element of wealth (or an argument of the Z-function) *per se*.

The characterization of all economic agents and transactions by a non-decreasing function is similar (but not identical) to the traditional utilitarian treatment of individuals in economics. However, firms have not previously been treated in this manner. The wealth function of a firm means that the firm also evaluates stocks of commodities and money in its possession (in terms of production and marketing possibilities) as do individual consumers. The evaluation is predicated on the assumption that the firm is capable of anticipating possible actions (and transactions) resulting in changes in future wealth, afforded by its existing stocks and transaction possibilities.

One can think of the balance sheet of a firm as if it were incorporated into or derivable from a well-behaved Z-function. This function permits managers to assess the current wealth of the firm. At first sight, one could argue that firms do not produce goods for themselves, hence the commodities they consume or produce have no intrinsic worth for them, except insofar as they represent sunk money costs and/or raw materials available to facilitate potential future production and sales. But the quantities of material stocks in hand *per se* imply nothing about the profitability of using or selling them. The latter depends on the firm's q-cycles and, market prices. So (to avoid losses), profitability depends on expectations about the future. The same point was made long ago by Irving Fisher, in a slightly different context, on the possible decisions of a firm not in equilibrium:

> Imagine that a firm faces a significant excess supply of some commodity, say tooth-paste. In disequilibrium, even if there is a considerable excess supply, the price of the toothpaste can remain positive. If it is sufficiently high and there is no common medium of exchange in terms of which to measure profits, the firm producing the toothpaste might consider that it makes positive profits in spite of the fact that it does not sell a single tube of toothpaste. (Fisher 1926, quoted by Cartelier 1990, pp. 40–41; our translation)

In the above paragraphs there is an implicit equation between utility and money. In the case of a firm, the equation seems entirely reasonable. The *raison d'être* for a firm is to make a profit. The most common macroeconomic interpretation today is to equate utility with aggregate consumption, defined as GDP less investment. This is supposed to be maximized over time under budget and other constraints, subject to an implicit conservation law analogous to the law of conservation of energy in rational mechanics.

Some leading economists are now calling for a return to 'real' utility (i.e. a rejection of national income as a proxy) (Albert and Hahnel 1990; Ng 1997; Kahneman et al. 1997; Lane 2000; Frey and Stutzer 2002; also Layard 2003). Second thoughts along these lines are more than welcome, although acceptable (quantifiable) measures of 'real utility' will undoubtedly be elusive.

4.2 MATHEMATICAL CONSEQUENCES OF AAL DECISIONS

Evidently the avoidance of avoidable loss (AAL rule) must hold for every separate decision within the q-cycle as well as for the cycle as a whole. Granted the rule does not guarantee that no losses ever occur. On the other hand, decisions violating the AAL rule are likely to result in losses and frequent violations will lead to the disappearance of the agent from the market. An axiomatic development now follows:

Axiom: All economic agents choose to avoid avoidable losses (they obey the AAL rule).

Theorem: There exists a scalar state function $Z(X)$, such that $dZ > 0$ whenever there is a change of state, via an AAL process.

Lemma: AAL introduces a complete ordering of the possible economic states of an agent in terms of wealth. Note that the existence of such an ordering depends upon the assumption of transitivity – hence path-independence – which means that a given state of wealth does not depend upon the sequence of actions and transactions by which it is achieved. This assumption is relatively safe for firms, at least in the short run. It means that quite different combinations of goods (or services) and money may correspond to the same degree of wealth.

However, in consumption decisions by individuals, path independence cannot always be assumed. For smokers, a cigarette after dinner means much more than a cigarette during or before the meal, so food followed by smoke has a different internal value than smoke followed by food. In other words, the combination of food plus smoke does not have the same value regardless of the order in which they are enjoyed. This results from interdependence between the two services. By the same token, a vacation at the beach in winter followed by a vacation at a ski resort in summer have much less value than the same combination taken in reverse order. In this case, the difference arises from external conditions (the seasons).

However, the possibility of a complete ordering is essentially the same condition generally assumed for consumer preferences in standard utility theory. Path-independence, in turn, means that wealth (or utility) is *measurable* in principle, and that optimization is theoretically possible. For convenience we use the terms 'better' and 'greater' interchangeably, while 'worse' corresponds to 'lesser'. These comparisons characterize the directionality of the ordering.

The AAL principle requires that economic agents must be able to compare their wealth (Z) before and after any contemplated action. Observing the behavior of real agents we can state that if an action took place, then the new state was *expected* to be better than the initial one (according to the ordering principle). If an action is declined, however, it does not mean that the expected result is worse (in terms of the ordering principle) although it may be. The AAL principle excludes only the possible actions leading to less favorable *expected* outcomes, while allowing a set of actions leading to better expected outcomes. A further rule (*decision strategy*) specifies the agent's actual decision.

Observing the decisions of a given agent, we can infer an ordering uniquely characteristic of that agent. In effect, if the action q takes place, then $X + q$ is not worse than X. If the agent has a possibility to choose an action represented by the vector q_i, it means he can choose to change his economic state, as indicated below:

$$X_i \Rightarrow X_i + q_i \tag{4.1}$$

Conversely, of course, the agent may decline to make the change. However any change that is consistent with the AAL principle must be preferable in terms of expected results, with respect to the elements of X. If agent selects the action q, then

$$X + q \geq X \tag{4.2}$$

We can infer that $X + q$ is 'better' than X, whence

$$X + q \succ X \tag{4.3}$$

This ordering conveys the irreversibility of economic actions. As noted, the AAL rule demands that the ordering relationship be *transitive*. Continuity criteria also preclude the possibility of any lexicographic evaluation scheme, which would, in any case, be unnatural for stocks of goods. The *reflexivity* condition follows from the definition.

For the sake of mathematical convenience, we can now assume that commodities are infinitely divisible. This assumption is only needed to permit the use of differentials. Otherwise the whole story holds for indivisible goods as well.

Let n_g be the number of goods in stock. One can now represent the various possible stockholdings of the agent as the points of an $(n_g + 1)$ dimensional Euclidean space. The benefits of different stock combinations are evaluated by the agent based on its past experience and accumulated knowledge. This is the source of its expectations regarding the future benefits of stocks. In mathematical terms it means that the agent can apply the AAL rule on every point in a subspace of $[R^{n_g+1}]$. It follows from the completeness of the ordering that every possible state is evaluable and comparable with all the other states.

The learning process can change the dimensionality of the Euclidean subspace. Goods offering no benefit to the agent may disappear, while different goods may appear. Nevertheless we can assume that such evolutionary changes occur on a time scale longer than the time scale for normal economic decisions. Thus, for purposes of description of economic decisions the subspace can be considered to be constant in time, and complete.

The remainder of the proof is trivial. We can apply Debreu's theorem (Debreu 1959; Candeal et al. 1998), namely:

Theorem: If an ordering ($\succ$) with the properties of *reflexivity*, *completeness*, *transitivity* and *continuity* is defined on the set X then it can be mapped onto a continuous function with the property that $C(X) > C(Y)$ if $X > Y$.

The original argument of Debreu was applied to prove the existence of the utility function. It is therefore tempting, at first to equate our $Z(X)$ function with the conventional utility function. However, recalling that X consists of stocks of goods and money, and noting further that the AAL principle differs from the standard utility maximization principle, it follows that Z does not have the same properties as utility U, although the two functions are obviously related. The major difference is that U is an integral over a vector field, defined over commodity space and subject to troublesome integrability conditions like path independence, whereas Z is a scalar from the outset. As mentioned before, we call Z the wealth function, for lack of a better and more precise term.

4.3 PROPERTIES OF THE Z-FUNCTION

As noted already, the non-decreasing economic wealth function introduced here plays a role similar to that of the neoclassical utility function, but in this form it has no direct counterpart in standard static theory. In a timeless, equilibrium environment the Z-function can be reduced to the utility function, but it can also be applied in a changing or non-equilibrium economy. The traditional static utility function is not explicitly dependent on stocks, of course. However, in the dynamic case the idea of increasing or decreasing some stock is necessarily explicit.

The usual problem is to allocate production or consumption optimally over time. For instance, many examples in the economics literature since Ramsey and Hotelling have assumed an initially fixed resource base that is used up (consumed) over some time horizon. The case we want to consider next is essentially an elaboration of the Hotelling model (Hotelling 1931), far from equilibrium, allowing for multiple goods, some durable and some consumable (plus money), over time. In principle, the dynamic optimization machinery is applicable here, at least in some simplified models. However, our computational procedure is rather different. First we explore the properties of the economic wealth function. We then derive an economic 'force law', and finally, based on it, the economic 'equations of motion'– or, more accurately, equations describing changes over time.

Every economic agent, whether a producer, a consumer (household) or a trader, has his/her/its own, unique economic wealth function. The Z-function implicitly contains all the relevant expectations (based on experience and knowledge) of the agent with respect to his/its economic activity concerning the future use of the stocks. When circumstances change an agent adapts in two ways. The first is through immediate stock adjustments (via trade or production) which change the numerical value of Z. The second, resulting from changed expectations, is through changes in the form of Z.

The mathematical form of the function, and the specifics of the argument (bundles of goods, plus money) will differ from agent to agent, as noted above. Several examples are given in Appendix C. To recapitulate again: the economic wealth of an economic agent is a function of the stocks of goods and money belonging to the economic agent:

$$Z = Z(X_1, \ldots X_N, M) \tag{4.4}$$

Thus a change in the Z-function is completely defined by changes in the stocks of goods and money. The mathematical form of Z expresses the agent's responses to changes. The equation

$$Z = \text{Constant} \qquad (4.5)$$

defines an N – 1 dimensional surface consisting of the indifference points, corresponding to complete closed q-cycles with no loss or gain. The sign convention is that $dZ > 0$ for allowed (loss-avoiding) processes, and $dZ < 0$ for forbidden (loss-making) transactions. This coincides with our intuitive notion of economic wealth. The economic wealth Z of an economic agent that obeys the AAL rule is therefore an increasing – on average – function of the stocks of goods and money belonging to the economic agent, and of time.

In loss-making q-cycles (of the first kind), or non-economic processes, Z will decrease, of course. However the AAL rule guarantees that the trend will be increasing. The Z-function is not completely determined by trade and production. Consumption beyond subsistence requirements is another way to decrease wealth. Theft and disaster losses are also possible, and when they occur they also decrease economic wealth.[2]

The most important properties of the economic wealth function can now be summarized (Ayres and Martinás 1996):

i.   Since economic wealth is a positive attribute, a function that measures economic wealth must be non-negative. Normally in trade and production processes $Z > 0$.[3]

ii.   Economic wealth comprises all goods and money, or money-equivalents (like receivables) that are owned outright (net of mortgages, debts or other encumbrances). The terms 'own', 'owned', 'ownership' etc. are shorthand for a more cumbersome phrase, such as 'to which the economic agent has enforceable exclusive access and rights of usage'.

iii. An increase in the agent's ownership of stocks of beneficial goods or money results, *ceteris paribus*, in an increase in the agent's economic wealth. In case of an incremental increase in the stock of a beneficial good (as opposed to a waste), where the stock of money is held constant, we can assert $dX > 0$, and $dZ > 0$. Similarly if $dM > 0$ and the stock of goods remains constant, then $dZ > 0$.

iv.   An economic agent's economic wealth can only increase or stay constant (but never decrease for significant periods) as a consequence of *economic* actions, that is, trade and production consistent with the AAL rule. In all other processes, such as the payment of taxes, consumption, wear and tear or obsolescence, Z decreases.

v.   Doubling all stocks will double economic wealth. This implies that the economic wealth function should have the property of homogeneity in the first degree. This is a useful property when it comes to selecting representative mathematical forms for Z, as noted in *Appendix C*.

vi.   The Z-function implicitly contains all information on the expectations
      of the agent. Its partial derivatives with respect to stocks of goods and
      money can be interpreted as *subjective* (or internal) value functions. It is
      understood that subjective values depend on the agent's existing stocks
      of other goods and money, as well as individual preferences. However
      the term 'subjective' is not to be understood as 'arbitrary' in any sense.
      For producers, the partial derivative of the Z-function is essentially the
      producer's internal valuation of the 'cost' of production. In the case of a
      firm, the cost is usually explicit, but the internal evaluation by a pro-
      ducer always includes some subjective elements (such as the choice of
      discount rates, depreciation rates and accounting methods) we can safely
      use the term 'subjective'. For individuals providing labor or other serv-
      ices, the cost may not be simply measurable in money terms.

The economic wealth function provides a new approach to the measurement of economic wealth of economic actors. In most real-time processes
(during which the agent does not change its expectations) Z has no explicit
time-dependence. Stocks may change from one time period to the next, of
course. All the short-term changes to the economic wealth function Z are due
to the changes of stocks; Z depends on time through the time-dependence of
stocks. There is an additional source of change over the longer term, namely
via knowledge accumulation, as reflected in changes in the form of Z. However we discuss the effects of knowledge later.

4.4   TIME DEPENDENCE OF Z: CHANGE AND GROWTH

We have noted previously that in our approach it is possible to address the
problem of change (i.e. growth) without the need to maximize an integral
over time, or to worry about satisfying integrability conditions. Assuming the
function Z is continuous and differentiable with respect to its arguments, and
assuming that the stock changes in any period are very small, we can write

$$Z(t + \Delta t) - Z(t) = \frac{\partial Z}{\partial X} \Delta X \qquad (4.6)$$

where

$$\Delta X = X(t + \Delta t) - X(t) \qquad (4.7)$$

We can further subdivide ΔX into two components, viz.

$$\Delta X = (\Delta X)_e + (\Delta X)_{ne} \qquad (4.8)$$

and

$$\frac{\partial Z}{\partial X}(\Delta X)_e \geq 0 \qquad (4.9)$$

where the subscript *e* refers to economic processes (i.e. trade and production decisions) consistent with the AAL rule. The subscript *ne* refers to non-economic processes, such as taxes or depreciation. For such processes it is almost always true (except in the case of receiving free gifts or subsidies) that the latter processes decrease material wealth, that is

$$\frac{\partial Z}{\partial X}(\Delta X)_{ne} \leq 0 \qquad (4.10)$$

Long-term survival as an economic agent requires that the net change (the sum of the two terms) must always be positive.

The partial derivative $\partial Z/\partial X_i$ can be interpreted as the marginal change in the Z-value of the ith good in the stock held by the agent. Thus, it defines the wealth change due to the stock change. We now introduce the following short-hand notation:

$$\frac{\partial Z}{\partial X_i} = w_i \qquad (4.11)$$

where w_i is the marginal Z-value of the ith good, which can be interpreted as the increase of economic wealth due to a unit increase of the quantity of ith stock. It is measured in wealth/quantity units. We emphasize that Z and w_i are characteristic of the agent. When necessary we will use the notation $w_{\alpha,i}$ denoting the marginal Z-value of the ith good to the αth agent. Similarly

$$\frac{\partial Z_\alpha}{\partial M_\alpha} = w_{\alpha,M} \qquad (4.12)$$

where $w_{\alpha,M}$ is the marginal Z-value of money to the αth agent. Marginal Z-values probably tend to decline with increasing Z, much like marginal utilities.

It is convenient here to introduce a new variable, as follows:

$$v_i = \frac{w_i}{w_M} \qquad (4.13)$$

which can be interpreted as the marginal value of the ith good, relative to the marginal value of money. With the above notation the differential change of economic wealth becomes:

$$dZ = \sum_i w_i dX_i + w_M dM \tag{4.14}$$

or, using *Equation 4.13*,

$$dZ = w_M \sum_i (v_i dX_i + dM) \tag{4.15}$$

Integrating *Equation 4.15* holding w_M constant and applying the homogeneous linearity condition yields a more transparent formula for Z, namely:

$$Z = w_M \left(\sum_i v_i X_i + M \right) \tag{4.16}$$

In this case the economic wealth of the agent is the sum of stocks weighted by marginal Z-values plus money weighted by its marginal Z-value. The colloquial notion of wealth (measured in monetary terms) is just Z/w_M. It should be pointed out that the simple solution *Equation 4.16* is not unique. There are more general solutions, discussed in Appendix C.

The economic wealth function of the agent depends on time only through the time-dependence of stocks. Disregarding locational factors, the change of Z for an agent α is given by the usual sum of partial derivatives

$$\frac{dZ_\alpha}{dt} = \sum_i \frac{\partial Z_\alpha}{\partial X_{\alpha,i}} \frac{dX_{\alpha,i}}{dt} \tag{4.17}$$

In the case of exchange processes, the stock change is the flow J plus applicable source terms, S

$$\frac{dX_{\alpha,i}}{dt} = J_{\alpha,i} + S_{\alpha,i} \tag{4.18}$$

Inserting the balance equations for stocks and money (Appendix B) we also obtain an equation in terms of the flow variables for the change of wealth of agent α, viz.

$$\frac{dZ_\alpha}{dt} = \sum_i \left[\frac{\partial Z_\alpha}{\partial X_{\alpha,i}} \left(\sum_\beta J_{\alpha\beta,i} + \sum_j y_\alpha^j T_{\alpha,i}^j \right) \right.$$
$$\left. + \frac{\partial Z_\alpha}{\partial M_\alpha} \left(-\sum_\beta p_{\alpha\beta,i} J_{\alpha\beta,i} + I_{\alpha\beta} \right) \right]. \tag{4.19}$$

where $J_{\alpha\beta,I}$ is the flow of the ith good from agent β to agent α, y_α^j is the level of production of the jth technology by agent α, $T_{\alpha i}^j$ is the technology coefficient of agent α which defines the quantity of the ith good used or produced in a unit production step. Finally, $p_{\alpha\beta,i}$ is price for the flow of the ith good from agent β to agent α. The term $I_{\alpha\beta}$ refers to money transfer unrelated to trade between agents α and β. *Equation 4.19* can include 'gifts of nature', if any, where 'nature' is defined as the $(n_a + 1)^{st}$ agent. However there is no Z-function for nature. See Appendix B for further details.

In this form the wealth equation is simultaneously a physical and economic expression. The J and y terms come from the physical balance equations for stocks and services, while the derivatives of the Z-function and the price contain the economic valuations. Investigating the wealth changes for strictly economic processes (trade and production) yields an economic interpretation of the derivatives.

4.5 SUBJECTIVE VALUES

The most surprising property of subjective value is that it is measurable, provided preference ordering does not change during the measurement period. Thence (in principle) subjective value is an empirical parameter. We have already noted that, for any good or service, the inequality $p > v$ is a condition such that the agent may be willing to sell. The higher the price the more willing the agent may be to sell. (It may be rational not to do so if there is a reasonable prospect of better offers, for instance). Similarly, under the AAL rule, the inequality $p < v$ guarantees that the agent will *not* sell. This feature of Z leads us toward a quantitative measurement of value.

In principle, there exists a limiting price p_0 such that the economic agent is indifferent between buying, holding or selling. (This price is sometimes called the *reservation* price.) The new measure does not apply only to markets in equilibrium. It is applicable in any pair-wise encounter between economic agents. The limiting price can be determined from non-equilibrium experiments.

Imagine a collection of N agents, indexed by $i = 1, \ldots N$. Suppose each is endowed with a different stock of goods and money, and a different set of initial expectations. Select the agent, indexed by α. Every agent offers to trade with others, one at a time (pair-wise), in some – possibly random – sequence. The agent with the lowest index number initiates the process (i.e. makes an offer to buy, or sell, a certain quantity of some commodity or service at some price). The offer may or may not be accepted, but some offers are accepted and some trades take place. The process continues until there are no more trades.

Note that although an agent who is confronted by an unpredictable sequence of possible trading opportunities cannot optimize in advance among possible outcomes, the end-result may come quite close to the optimum result, depending on the agent's trading strategy.

Now imagine a large number of replicas of this population of agents, differing only in terms of the sequence of interactions. Each trade increases the economic wealth of both agents and reduces the number of potential trades that can benefit both parties. Assume all the agents in the trade act in accordance with the AAL rule. Now select a typical agent, α. Monitor this agent's transactions for a selected good and plot them in price-quantity space. We would expect to see a clear pattern, such that smaller and smaller quantities exchanged cluster around a 'zero' price-quantity point. This point corresponds to the indifference point.

In the absence of the AAL rule one would expect to see no correlation between the offered price and the quantity exchanged. In our *gedanken* experiment there is also a maximum offer price p_0 at which some agent will agree to buy the good, and a minimum price p_s at which the agent will sell. The inequality $p_0 < p_s$ always holds. As the number of experiments increases, the difference between the two prices will gradually disappear.

We can now formalize a definition of value: for any agent the internal or subjective value of the ith good is v, where $p_0 < v < p_s$. Repeating the above experiment with other cases where the agent begins with differing stocks, one can experimentally determine the subjective value as a function of the size of the stock, viz.

$$v = v(X, M) \qquad (4.20)$$

In summary, the subjective value of a good is defined for (and by) each agent. Moreover, it is known only to the agent. Note that while some agents, notably producers, may determine this value – a target selling price – based on objective production costs plus a fixed markup, there is no need for them to do this, and some agents do not. Consumers obviously do not fix subjective values of the basis of cost of production (which they are unlikely to know) although budgetary constraints are likely to have a strong influence. It is important to note that, since each agent may have a different decision rule, two agents would not be likely to assign the same value for each good, even if both agents were momentarily in the same economic state in terms of stocks of goods and money.

4.6 SUBJECTIVE VALUE OF MONEY

If the agent spends an increment dM of money, with no change of its stocks of goods, as for instance when an agent pays taxes or purchases a service that cannot be stockpiled, then its economic wealth decreases by an increment

$$dZ = w_M dM \tag{4.21}$$

(Needless to say if the agent receives a tax refund, a gift, or an inheritance, which is not a payment for services, dM will be greater than zero.) The variable w_M cannot be determined on the basis of simple exchanges, at a point in time, though there is a time preference (discount rate) to take into account with respect to exchanges of money at one time for money at another time.[4] There are also variable exchange rates between different currencies. More precisely in normal exchanges, when the good is exchanged for money, only the relative value of the good appears, namely w_i/w_M.

Differentiability of Z implies the symmetry of the second partial derivatives of the Z-function, viz.

$$\frac{\partial w_M}{\partial X_i} = \frac{\partial w_i}{\partial M} \tag{4.22}$$

Integrating the above relation, we see that the change of w_M between two states is given by the formula

$$w_M = \int \sum_i \left(\frac{\partial w_M}{\partial X_i} \right) dX_i + \int \left(\frac{\partial w_M}{\partial M} \right) dM \tag{4.23}$$

Assuming, further, that Z is a linear homogeneous function of its arguments (as noted above), it follows that w_M must also satisfy the equation

$$dw_M = -\sum_i \frac{X_i}{M} dw_i \tag{4.24}$$

Taking into account the values, $w_i = v_i w_M$, we obtain

$$\frac{dw_M}{w_M} = -\sum_i \left(\frac{X_i}{M} \right) dv_i - \sum_i \left(v_i \frac{X_i}{M} \right) \frac{dw_M}{w_M} \tag{4.25}$$

whence

$$w_M = w_0 \exp\left(-\int \sum_i \frac{X_i dv_i}{M + \sum_k X_k v_k} \right) \tag{4.26}$$

The above equation fully determines the subjective value of money, w_M, except for a constant multiplier.

We are free to choose the constant multiplier w_0 for convenience. All values $w_0 > 0$ are acceptable. It means that the measuring unit of economic wealth is defined only by the agent, and for normal (individualistic) agents that parameter is inaccessible to others. The subjective economic wealth function is now fully specified, viz .

$$Z = w_0 \left(\int w_M \sum_i v_i dX_i + \int w_M dM \right) + Z_0 \tag{4.27}$$

The two parameters w_0 and Z_0 are important to the agent but they play no role in the economic interactions between agents. The economic wealth function of an agent can be inferred, in principle, by observation of (a large enough number of) transactions. However this is seldom, if ever, practicable.

The Z-value of money does not appear in real economic processes. Perhaps it should. Other factors remaining the same, a money transfer from an agent with a smaller subjective value of money to an agent with a higher subjective value of money increases the total welfare of the two taken together. A comment attributed to Jeremy Bentham makes the point: 'By a particle of wealth, if added to the wealth of him who has least, more happiness will be produced, than if added to the wealth of him who has most'. This is another way of saying that wealth has declining marginal utility.[5]

A few words on the economic interpretation of Z seem called for at this point. Abstract theory is all very well but most people, ourselves included, handle theory better if there is a straightforward point of contact with the 'real' world. We have used the term 'wealth function' although we originally labeled Z as 'economic entropy' (Martinás 1989; Bródy et al. 1985), then in a later version we called it a 'progress function' (Ayres 1994, chapter 7, pp. 163–73. In a still later iteration of this work we tried 'welfare' before finally settling on 'wealth'. The point is that our use of the word is not necessarily identical to normal (rather imprecise) usage. But, on the whole, it seems better to associate the Z-function with a familiar word than to invent a new one.

Having said this, it is important to note that the Z-function, as such, is not measured in monetary units: more precisely Z is the product of monetary wealth times the value of money, defined in this section (*Equation 4.16*). We can now go further and suggest possible measures of 'real' wealth that could be used as proxies for Z in future empirical studies. It seems clear that 'real' wealth, in practice, is always measured in financial terms, even though we have insisted that material goods may constitute an important component. But the value of goods to an agent in business may be very different (and much higher) than the value to others. When a bankrupt firm is liquidated, for instance, the goods in stock are simply auctioned off, often at considerable loss. A liquidated firm is likely to be worth much less than a going concern. Similarly, when an estate is probated and valued for tax purposes, the artworks and furnishings are counted at current market value, as determined (perhaps) by an auction or professional assessor. However, to the last owner of the property, the subjective value of the items is likely to have been much higher.

In a world of perfect markets, the best proxy for the real financial value of a firm that is not bankrupt – for instance as a target of acquisition – should be determined by the market value of its shares. However, when the firm owns land, mineral reserves, patents, or shares in other firms, or when it has unrecognized liabilities, the market price of shares can be very misleading (which accounts for the spectacular 'coups' of some successful corporate raiders in the 1980s). However, these are problems for the stock market analyst or investment banker.

But the question remains: for purposes of future empirical studies: can we equate the Z-function with objectively determined wealth, in monetary terms? In the case of a firm, the answer is probably 'yes' but with some caution. The reason is that the subjective (internal) and objective valuations in this case should not be grossly different. In the case of individuals or small businesses, the monetization of wealth is much less straightforward, and the internal valuation of stocks of goods for a going concern is likely to be very different from the value to others. However, we cannot pursue this interesting topic further in this book.

4.7   ON WEALTH MAXIMIZATION

In conventional economics it is traditional to maximize an integral of net consumption over time, subject to budgetary and other constraints. In such exercises it is traditional to assume a constant rate of growth, driven by exogenous improvements in productivity, while the economy remains in equi-librium. Even under such simplistic (and unrealistic) conditions, constrained maximization of an integral over time requires sophisticated mathematical techniques, usually by means of optimal control theory (e.g. Arrow and Intriligator 1981; Pontryagin 1962). A rather simpler approach, widely adopted in economics, involves the use of a Hamiltonian function, which implicitly assumes the existence of an underlying conservation law or rule analogous to conservation of energy (Mirowski 1984).

Evidently any agent (firm) that knows its own Z-function can analyze the parametric conditions for short-term wealth maximization, or growth rate maximization, by means of ordinary calculus, without the difficulties associ-ated with maximizing an integral over time, and without assuming any underlying conservation law. As explained above, Z is a function of the economic state of an agent, being additive but not conserved. The expected wealth increase due to a possible action is the driving force for undertaking that action. The simplest strategy for the agent is to choose that action which leads to the short-term (myopic) maximization of Z. This is a constrained maximization, where the applicable conservation laws are mass balance (de-rived from physics) and conservation of money in every transaction. In Chapter 5 it will be shown that myopic maximization in a changing environment is generally not the best strategy for long-term wealth maximization.

The special case of price-taking firms operating in or near equilibrium conditions, where instant wealth can be measured in objective monetary terms, may be of some practical interest to accountants. However, in the real economy, financial analysts are primarily interested in the value of the firm as a whole, which in practice is largely attributable to the stock market's valuation of future profit potential. In reality the present and future profitability of any firm de-pends upon the interplay of the firm with other firms, in a changing and non-equilibrium environment. In Chapter 5, which follows, we explore the possible decision rules for an agent acting in a non-equilibrium environment. In Chapter 6 we examine the effect of interactions between firms or agents in a non-equilibrium environment. We also consider the stability of equilibria, un-der different initial conditions and various perturbing influences.

NOTES

1. A formal mathematical proof of this proposition has been given elsewhere (Martinás 1996). The discussion in this chapter is more informal and correspondingly somewhat sketchy.
2. Taxes are not easy to characterize. Some taxes can be regarded as payment for government services. On the other hand the tax paid is not directly related to the service provided, whence some taxpayers subsidize others. Those who pay more than they receive in services could regard the difference as a loss, and conversely.

3. True, it is possible to have negative wealth (i.e. debts not covered by assets), and this occasionally occurs, primarily as a result of misuse of credit cards. Credit card companies have justified widespread creation of unsecured debt by relying on potential future earnings in lieu of assets. This is a very complex subject that is not worth detailed analysis at this stage of the development of our theory.

4. Gold formerly served as 'real' money, and there is still a group of people who believe that gold is a 'store of value' more enduring than paper money. Whatever the theoretical validity of this belief, gold has a monetary value that fluctuates according to production, competitive non-monetary uses of the metal, inflation and investor opinion.

5. An interesting, if peripheral, application of the above derivation is the following. We can suppose a dictator (who is not familiar with the extensive literature on how social wealth is defined and how it is related to the individual wealth of the citizens) makes the simplistic assumption that the total wealth of society is the simple sum of the Z-functions of all its individual citizens. He wants to tax his society in such a way as to increase his own wealth without adversely affecting the total economic wealth of the rest of the country, as understood by him. He divides his citizens into two classes (they could be 'landowners' and 'workers') such that each member of a class shares the same Z-values. The two values are Z_1 and Z_2. The total wealth of the rest of the dictator's society is

$$Z = Z_1 + Z_2 \qquad (4.28)$$

If the dictator confiscates (i.e. taxes) an amount of money m_1 from each member of the landowner class and gives an amount m_2 to each member of the worker class, then the total per capita economic wealth of the rest of society changes by an amount

$$w_2 m_2 - w_1 m_1 = w_2 \left(m_2 - \frac{w_1}{w_2 m_1} \right) \qquad (4.29)$$

The total economic wealth of the rest of society remains constant if the dictator imposes a money tax (or rebate) on each individual as follows

$$m_1 - m_2 = m_1 \left(1 - \frac{w_1}{w_2} \right) \qquad (4.30)$$

In this case the efficiency of taxation can be characterized as

$$\eta = 1 - \frac{w_1}{w_2} \qquad (4.31)$$

In this case the constant total wealth relation defines the constants for the two classes of agents and only the overall (system wide) value of money remains to be selected.

For those who know thermodynamics there is a somewhat misleading similarity between equation 4.31 and the Carnot efficiency formula. This suggests a possible interpretation of Z as the economic 'temperature'. However we think this analogy is probably a stretch too far.

5. Decision-making strategies

5.1 INTRODUCTION

As discussed in Chapter 4, each agent has a unique wealth function, Z. This function reflects all the relevant economic information about the agent's stocks of goods and its expectations concerning potential stock changes. It also reflects the agent's expectations with regard to the future behavior of other agents with which it normally interacts, based on past experience. Agents – or outside observers –can learn about other agents' preferences and expectations by inference from their actions in various circumstances. This information enables the agent to decide whether a possible transaction is consistent, or not, with the expected AAL rule.

In the present chapter we investigate exchange mechanisms in several kinds of markets. There are several important distinctions to be made. First, individuals and small businesses are normally *price-takers*. They may not be willing, or even allowed, to bargain. Most retail markets nowadays do not encourage bargaining. At the opposite extreme, there are a few (but important) examples of buyers – often government agencies (such as the US Defense Department), large commodity traders (like Cargill), major manufacturers (like Ford or GM) or giant retail chains (like Wal-Mart or Sears Roebuck) – that are *price-makers*, because of their ability to select among many competing would-be suppliers. Finally, there is an in-between category, consisting of artisans, larger shops and medium-sized manufacturers that may act like price-makers with respect to smaller consumers and price-takers when dealing with giant customers, but who can, and do, bargain when dealing with each other.

Buyers in most retail markets may or may not be willing to buy variable quantities, depending on price. The most common case, of course, is where the buyer is looking for one unit or a fixed quantity. However there are some large buyers – such as professional traders, wholesalers and investors –who are willing to bargain over both price and quantity simultaneously. Sellers, on the other hand, are normally willing to bargain for both quantity and price at the same time, or separately. As noted above, sellers may, or may not, be price-makers, depending on their market power, but even so they are often willing to offer quantity discounts.

Another important distinction is between individual (pair-wise) bargaining and auctions. In auctions, quantities are fixed and many buyers bid for a given item until the highest bidder obtains it. In pair-wise negotiation between a single buyer and a single seller, we can assume that neither party is either a price-taker or a price-maker. Both quantity and price may be negotiated, but by far the usual case is price negotiation for a fixed quantity. The major example is the real estate housing market for homes.

A third important distinction is strategic. The strategy assumed in most economic textbooks is *myopic maximization*, where each party tries to maximize its short-term gains. However there are many cases where this strategy is not optimal because of unavoidable uncertainty (as will be shown below) and where the best strategy – in terms of maximizing the Z-function – is what we call *adaptive*. The AAL rule provides a governing principle for the adaptive decision rule, as will be shown. There are some cases where an apparently random decision strategy in the face of uncertainty might be adopted, though it would seldom, if ever, be optimal.

The chapter concludes with a brief discussion of production-related decisions unrelated to trade, namely decisions regarding the level of production and the choice of technology, at the micro-scale.

5.2 THE MOST GENERAL CASE

We now consider an exchange economy, involving n agents, exchanging k goods. As always, an exchange consistent with the AAL rule implies that each agent's economic wealth increases as a result of every trade. The wealth function of the agent depends on time, but only through the time dependence of stocks. The change of Z for an agent α as a result of stock changes is given by the usual sum of partial derivatives

$$\frac{dZ_\alpha}{dt} = \sum_i \frac{\partial Z_\alpha}{\partial X_{\alpha,i}} \frac{dX_{\alpha,i}}{dt} \tag{5.1}$$

If $J_{\beta\alpha,I}$ is the flow of stock I from agent β to agent α, then in general,

$$\frac{dX_{\alpha,i}}{dt} = J_{\alpha,i} \tag{5.2}$$

and

$$J_{\alpha,i} = \sum_\beta J_{\alpha\beta,i} \tag{5.3}$$

In Chapter 4 we introduced the following short-hand notation:

$$\frac{\partial Z_\alpha}{\partial X_{\alpha,i}} = w_{\alpha,i} \tag{5.4}$$

where $w_{\alpha,i}$ is the marginal Z-value of the ith to the αth agent. Similarly we defined

$$\frac{\partial Z_\alpha}{\partial M_\alpha} = w_{\alpha,M} \tag{5.5}$$

where $w_{\alpha,m}$ is the marginal Z-value of money to the αth agent. Finally, we let

$$\frac{w_{\alpha,i}}{w_{\alpha,M}} = v_{\alpha,i} \tag{5.6}$$

and the ratio $v_{\alpha,i}$ can be interpreted as the marginal Z-value (expressed in monetary units) of the ith good to the αth agent. Hereafter we simply call it *value*. An exchange consistent with the (expected) AAL rule implies that the agent's wealth increases. Rewriting *Equation 5.1* using the short-hand notation we obtain

$$\frac{dZ_\alpha}{dt} = w_{\alpha,M} \sum_{i\beta} (v_{\alpha,i} - p_{\alpha\beta,i}) J_{\alpha\beta,i} > 0 \tag{5.7}$$

From the inequality $\frac{dZ}{dt} > 0$ and the assumption that $w_{\alpha,M} > 0$ it follows that

$$\sum_{i\beta} (v_i - p_{\alpha\beta,i}) J_{\alpha\beta,i} > 0 \tag{5.8}$$

must hold. If we consider only transactions involving the ith good then the inequality must also hold for each of the individual terms in the sum, viz.

$$(v_i - p_{\beta,i}) J_{\beta,i} > 0 \tag{5.9}$$

To satisfy this inequality either both terms must be positive or both must be negative. Thus, if $v_i - p_{\beta,i} > 0$ then $J_{\beta,i} > 0$. This means that agent α is willing to buy from agent β. In the opposite case, if $v_i - p_{\beta,i} < 0$, then $J_{\beta,i} < 0$, which means that agent α may be a seller and agent β is a potential buyer. Hereafter we can omit the indices α and β.

This justifies our interpretation of v_i as the subjective value of the ith good to the agent. In words, if the subjective value of the good to the agent is greater than the price being offered, the agent may or may not buy, but will not sell. If the subjective value to the agent is less than the price offered, the agent may or may not sell but will not buy. Either way, the transaction is economically irreversible in the sense that it can only proceed in one of the two directions.

Equation 5.7 is normally valid only for the exchange of a small quantity of the good, as the subjective value v also depends on the quantity of stocks already held by the agent. If the agent buys more of the good, its money stock decreases and, in most cases, the marginal (subjective) value of additional increments of the good also decreases. (This is essentially the well-known 'law' of declining marginal utility, discussed in Chapter 3 and at the very end of Chapter 4.) The same logic holds, in reverse, for a decision to sell.

For agents characterized by declining marginal (subjective) value of increasing stocks, it follows that there is an upper limit to the quantity that can be traded at a given price. (The lower limit is zero.) The wealth change resulting from a purchase must be a hyperbolic – inverted U shaped – function of the quantity bought or sold. It increases from zero when there is no trade to a maximum, when $v = p$, and declines again to zero as the quantity exchanged increases still further. The maximum quantity that can be exchanged, J_{max}, can also be determined in the same way from condition $\Delta Z = 0$.

In a market consisting of many price-takers interacting pair-wise, trading stops when the subjective value of each agent reaches an indifference point. However, the outcome can easily depend upon the order in which the encounters between potential trading partners takes place. A sequence of trades in order of increasing profitability yields a different outcome than a sequence in the reverse order, unless the agent is blessed with perfect knowledge about each trading partner. $\Delta Z = 0$ defines the lower and upper limit to supply (demand) at a given price. All values inside the limits are in agreement with the AAL rule.

5.3   DECISION STRATEGIES IN AN UNCERTAIN WORLD

There are many possible strategies for traders, but two examples are worthy of discussion. They are as follows:

(1) *Myopic maximization*   The agent ignores future possibilities and seeks to solve the instantaneous maximization problem,

$$v\left(X + J, M - \sum pJ\right) = p \qquad (5.10)$$

The corresponding decision rule is simply to select the quantity J_m to be sold or bought, such that the new subjective value v after each sale (purchase) will be equal to the price offered. The traded quantity J at which the trading gain is maximized can be determined graphically or analytically if the mathematical form of v as a function of X and M is known. In the (unlikely) case that the agent has perfect information this must be the optimum strategy.

(2) *Adaptive proportional coupling*   The size of each sale or purchase is a function of the expected gain resulting from it. The simplest and most convenient function is the linear relationship, where the quantity J to be sold to or bought from any given customer depends upon the value–price differential, according to the simple rule

$$J = L(p - v) \qquad (5.11)$$

where L (in this case) is a parameter, to be determined on the basis of experiments or experience. It appears that there is an optimal (gain maximizing) value for L, which must also be determined from experience or by experiments. On reflection it is obvious that J could be many functions of the difference between price and subjective value. For instance, it could be an exponential function or a power law. Evidently the parameter L couples the two. It expresses the expectations of the agent on the expected possibilities of exchanges, given the dynamics of the market. It is adaptive, in the sense that the optimal value for L has to be learned in any given market, and can change as conditions change. In the most general case, an agent may have different couplings for different trading partners and different goods.[1] Furthermore, there can be cross-effects, for example, the trading of one good may increase or decrease the trading of other goods. More generally, J, p and v are vectors. This adaptive coupling term is therefore a tensor, the elements of which take the general form:

$$L = L_{\alpha\beta,ik} \qquad (5.12)$$

where α is the agent willing to make the exchange, β is the index for the partner, i is for the exchanged good and k is the index for the other exchanged goods. A more complete discussion of the properties of L is provided in Appendix D.

It is convenient for purposes of discussion and interpretation to think of *Equation 5.11* as an economic 'force law' where $p - v$ is the driving force and the flow J is the resulting effect. The price–value differential on the right-hand side of *Equation 5.11* represents the strength of the incentives to trade, while the coupling tensor L defines the magnitude of the response for a unit driving force,[2] taking into account the agent's expectations and learned behavior. It follows that L is a generalized version of the demand curve in standard textbooks.

5.4   AUCTION MARKETS AND MAXIMIZATION

An auction is a method of price maximization or cost minimization for a fixed quantity of a good. Auctions are most frequently held for unique items, such as artwork or other collectibles, or foreclosed real estate, although petroleum drilling sites or permits for TV stations can and sometimes are auctioned. The maximization principle, together subject to the supply-equals-demand equilibrium condition, defines a unique solution. Assuming $J_{\alpha i}$ is the quantity of the ith good that the agent α buys for price p_i, the equilibrium condition is defined by *Equation 5.11* where the sum over all physical flows vanishes (due to conservation of mass)

$$\sum_{\alpha} J_{\alpha i} = 0 \qquad (5.13)$$

and $J_{\alpha,i} > 0$ refers to a buy, while $J_{\alpha,i} < 0$ refers to a sale.

In an auction market all of these equations (for each index number) must be satisfied simultaneously and there is a single price p_i for the ith good, at which aggregate supply and aggregate demand of that good are equal. This is the equilibrium condition. Once the Z-functions of all agents are determined, everything is determined.

These equations are similar to traditional general equilibrium conditions, except that in the standard microeconomic theory agents are usually required to spend all their money on goods, whereas in our case economic agents also attach subjective values to money *per se*. They decide on the quantity of money to be spent (subject, of course, to the quantity available). The other difference is that, in contrast to standard theory, our equations are not homogeneous first order in prices. This is due to the fact that money is built into the wealth function.

5.5   AUCTION MARKETS, ADAPTIVE STRATEGY

In this case, from *Equation 5.11*, the price is defined by the total supply equals total demand condition for each good, whence

$$\sum_{\alpha} J_{\alpha,i} = \sum_{\alpha} L_{\alpha,i}(v_{\alpha,i} - p_i) = 0 \qquad (5.14)$$

Solving *Equation 5.15*, we obtain the equilibrium price for the ith good

$$p_i = \frac{\sum_{\alpha} L_{\alpha,i} v_{\alpha,i}}{\sum_{\alpha} L_{\alpha,i}} \qquad (5.15)$$

5.6 PAIR-WISE EXCHANGES

Pair-wise bargaining between agents that are neither price-makers nor price takers is also described by the above rules (*Equations 5.11* or *5.14*), for myopic maximizers and adaptive couplers, respectively. For myopic maximizers, *Equation 5.11* specifies the traded quantities in each encounter. For the adaptive agents, *Equation 5.14* defines the quantity traded, which is generally different from the myopic case. If one of the agents is a price-maker, of course, there is no real bargaining process and the price-taker can only decide how much of the good (if any) it wants to buy or sell at the given price.

However, in pair-wise bargaining the outcome depends on the order in which the agents encounter each other. If the most favorable offers are encountered first, the trader will end up far better off than in the reverse case. Of course, the more realistic situation is that the encounters are random (or, the order may be based on parameters unknown to the agent) and it is precisely the strategy for dealing with this uncertainty that determines the outcome. The example which follows illustrates this point.

5.7 THE BAKER'S DILEMMA

To clarify some of the strategic choices discussed in abstract terms above, it may be helpful to introduce an example. This example is rather specialized but it does distinguish the various strategic possibilities. Imagine a baker who produces 100 loaves of bread each morning. Whatever remains unsold at the end of the day is consumed by his family. His profits are used to buy supplies for the next day's baking and to buy other goods for his family. The initial stocks at the opening of business in the morning were set at 100 loaves of bread, and $200. The baker's starting subjective value of his bread is $2 per loaf.[3] The number of would-be buyers was selected from the range 50 < N <

150, offering prices from the range $2< p < $10, by a random number generator.

To investigate the different strategies available to our hypothetical baker we selected an explicit Z-function, as follows

$$Z = \sqrt{XM} \qquad (5.16)$$

where X is the quantity of bread in stock at any moment and M is the amount of money in the till.

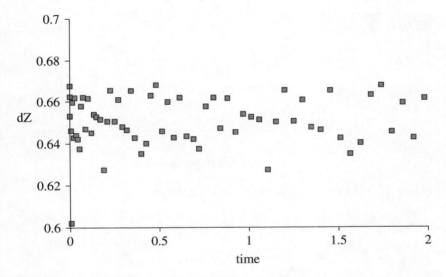

Figure 5.1   Wealth gains for myopic maximizing strategy

For each strategy the baker's sales can be calculated by solving *Equations 5.11* and *5.14*. The results are plotted for different strategies in *Figures 5.1, 5.2, 5.3*. For purposes of exposition it is convenient to define the maximum gain the baker could achieve in theory, that is, if his customers arrived in strict order of their price offers, highest offers being first in line. The same result would be possible if the baker knew the subjective utilities of each of his customers. We call this 'cream skimming' limit 100%. Not having this information, of course, the baker can only make his quantity-per-customer decisions based on chosen strategy. The results are plotted as averages over ten days with buyers arriving in random order. The simulation results show that, among the three strategies considered, the proportional strategy is better than the myopic maximizing strategy.

Figure 5.1 shows the wealth gains for the perfectly rational 'maximizing' baker over a period of time. Where the buyers arrive on a random basis this trader averages 65% of the maximum possible cream-skimming gain.

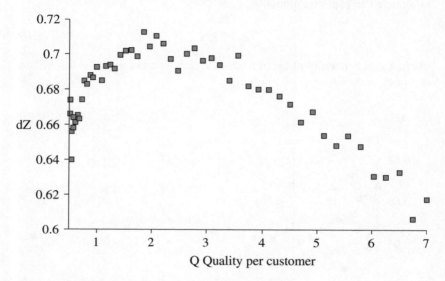

Figure 5.2   Wealth gains for the randomizing strategy

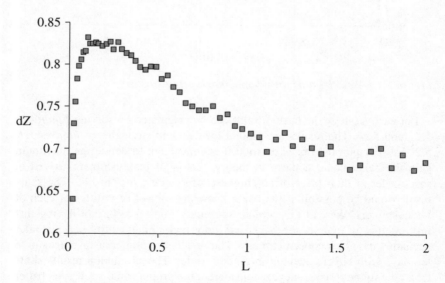

Figure 5.3   Wealth gains for the adaptive coupling strategy

Real trade decisions may depend on parameters not identified or not understood. *Figure 5.2* shows the results when the trader adopts the second strategy. In this case, the best strategy for an agent (lacking better data) may be a random choice of J in the interval between zero and J_{max}. Again, we simulated trading over 10 days, with the same randomly chosen set of buyers on each day as in *Figure 5.1*. However, in this case the baker's profits depend on his choice of the parameter q. If q is chosen to be too small, the baker will end the day with a lot of unsold bread. On the other hand, if q is too large, the baker will sell out too quickly and miss too many of the customers willing to offer higher prices. (For very large q the baker's gains can actually be negative, although this would be very unlikely in reality.) Thus the baker's profit in this case is an inverted U-shaped function of q, with a maximum. But surprisingly, the maximum gain in the simulation is 70%, of the cream-skimming maximum, considerably better than the gain achievable by the 'rational maximizing' agent. The best value of q is a function of the size of the initial stock and the number of expected buyers. In our simulation it turned out to be about 5% of the trader's starting stock.

Figure 5.3 shows the simulation results achievable by adaptive coupling strategy, as a function of the parameter L. In this case each choice of L was tested for an iteration consisting of a complete set of 10 trading days and the same set of customers arriving in each day as in the two previous simulations. As we suspected, there is indeed a best value (or range of values) for L which, in our example, yields a gain by the baker of more than 82% compared to the theoretical ('cream-skimming') maximum.

It can be argued that each simulation involves a rather special case, with little relevance to the real world. (More relevance, nevertheless, than the bakers in elementary economics textbooks, who miraculously produce bread from rented capital and labor services alone (e.g. Mankiw 1997).)

5.8 PRODUCTION-RELATED DECISIONS

We now recapitulate the discussion of production of goods or services by a firm, in Chapter 3. Production involves what we have termed q-cycles, beginning with stocks of goods and money in inventory and ending when the original stocks of goods are replaced and the stock of money is (normally) larger than it was at the beginning. Production is a transformation. All the necessary machines, raw materials and intermediates are stockpiled.

Within the q-cycle there are voluntary purchases of raw materials and sales of finished products or services, as well as involuntary expenses, including capital amortization, maintenance of equipment, labor, fuels, utility services and (in the consumer's case) food. The voluntary purchases and sales by producers are trades subject to the same decision rules as other trades, except for the subjective values in *Equation 5.7*. This case is normally thought of as the known costs of doing business, with very little preference component. These known costs include the involuntary items mentioned above, as well as margins for market risk and desired profit levels, as determined by the entrepreneur.

Production decisions within the q-cycle are of several kinds. The simplest is whether to produce or not to produce, depending on the state of the market. However a short-term decision not to produce is comparatively rare in modern industry. (If the industry has significant overcapacity, the likely result is that one or more plants will be closed.)

Another theoretically possible choice facing some producers, within the q-cycle, is the choice of technology, depending perhaps on the availability of labor *vis-à-vis* other inputs. Again, this choice is often trivialized in economics texts by confusing it with irreversible choices involving long-term investment in capital equipment, reorganization and training. The latter is an investment, at least in part, in knowledge. The time scale of such decisions encompasses many q-cycles. While the behavior of a large group of producers, some entering and some leaving a given market, may appear from a distance to be similar to the behavior of a group making choices in real time, the two situations are not really equivalent.

5.9   PRODUCTION DECISIONS UNRELATED TO TRADE

A third sort of choice within the q-cycle involves the simultaneous selection of product – or product mix – and level of production, by multi-product firms. Such decisions are applicable to a significant group of real producers. They are not usually made on a daily basis, but it is possible (e.g. for a machine shop) and the frequency of change depends on the batch size, which is characteristic of the industry. These micro-decisions can be expressed in mathematical language, but it is sufficient for our purposes to note that managers in the real world seldom think in such terms, still less do they attempt to optimize. They do, however, use rules-of-thumb which incorporate AAL conditions.

In Appendix B we derive the equation for stock changes in a production process as follows:

$$\Delta X_i = yT_i + J_i \qquad (5.17)$$

where T_i is the technology vector normalized for a unit product; y is the level of production (quantity of product/unit time). As mentioned previously, the use of a purchased service – such as electric power, water, sewage, or insurance – is necessarily accompanied by a trade. In *Equation 5.18*, J_i is the quantity of the ith service bought at price p_i. Hence the money stock of the producer also changes during production:

$$\Delta M = \sum_S p_S y T_S \tag{5.18}$$

Inserting both of these relationships yields

$$Z(y) = Z(X_1 + yT_1, \ldots, X_n + yT_n, M - \sum_{n+1}^{n+s} yp_iT_i) \tag{5.19}$$

Then

$$dZ = Z(y) - Z(0) \tag{5.20}$$

is the expected wealth gain (value-added) by the production process *per se*, not taking into account unavoidable costs of being in business – accounted for elsewhere – nor trade associated decisions, such as the purchase of raw materials, hiring of extra production workers and the sale of finished goods or services.

The AAL loss rule demands that $0 < y < y_{max}$, where y_{max} is the solution of the equation $Z(y_{max}) = Z(0)$. The possible strategic decision rules for the level of activity y include myopic maximization on a case by case basis and adaptive coupling (to value-added) analogous to the decision rules for trading. A random choice of activity level is also possible. We note that the levels of production are also capacity limited, for example, $0 < y < y_u$ where y_u is the maximum physical production capacity for a firm.

Myopic maximization   The agent selects y_o such that $Z(y_o)$ is maximized.

Adaptive coupling   The agent expects a wealth gain (value-added) from production as follows:

$$F = \sum_i^n v_iT_i - \sum_{i=n+1}^{n+s} p_iT_i \tag{5.21}$$

Here F can be considered as the anticipated wealth increase from a unit output of the product, at the production stage. In this case index $i = 1 \ldots n$ covers the stocks, and $i = n + 1, \ldots, n + s$ covers the services. F *is* the 'driving force' for production. In the case where the agent has a single technology, the choice is simply

$$y = LF \tag{5.22}$$

The term L is the coupling parameter. This parameter is different for different agents and technologies, so the general form of coupling for agent α, using the jth technology is L_α^j.

Random   The agent's choice might depend on factors not encountered in the present description, so it appears as a random choice in the $0 < y < y_{max}$ interval.

NOTES

1. In principle, agents may prefer to trade with some partners and not with others. In practice, this sort of discrimination might result from experience, or reputation, although other factors could also enter. However, we do not consider preferences of this sort hereafter.
2. In this form the trade relationship strongly resembles Ohm's Law in physics, where the current flow is proportional to the voltage, and the linear parameter is the electrical conductivity (the inverse of resistance), which varies according to the conductor. A similar relationship relates heat flow to material characteristics in thermodynamics by Onsager (1931). Again, the coefficients reflect the thermal conductivity (or resistance) which are properties of materials and must be determined by experiment.
3. Recall the comment in the last chapter about subjective valuation. The baker is quite likely to set his target price for bread on the basis of costs plus a desired profit, but other factors are involved as well.

6.   Dynamics

6.1   INTRODUCTORY REMARKS

When an economic agent experiences wealth loss, for any reason that is not purely accidental – an 'act of God' – it has to change its economic behavior to avoid repeating the same behavior and making the same mistakes. Learning and adaptation, in this sense, are necessary to avoid extinction (i.e. disappearance from the economic system). The simplest response by a firm is simply to change its stocks of goods, whether commodities – if the firm is a trader – or perhaps replacing obsolete capital equipment, in the case of a manufacturer. The latter would also necessitate a corresponding change in procedures and labor skills needed.

A second adaptation would be to modify the rules-of-thumb the firm uses to make decisions, notably by modifying its decision rules with respect to pricing and production levels (in the case of a manufacturer). This is a straightforward kind of learning. These adaptations are expressed in the L-tensor. Recall that L summarizes the expectations of the agent with respect to future possibilities for buying, selling or production. Roughly speaking, it corresponds to *expected demand*.

A third adaptation – often undertaken out of desperation when other adaptations prove insufficient – is to undertake a radical innovation of some sort. We discuss radical innovations later. The remainder of this chapter deals only with dynamics and incremental technical change in the sense of learning-by-doing on the part of both workers, managers and designers.

If the agent is a trader who is a price taker with only one possible trade partner, the best strategy is probably to maximize the volume of trade. On the other hand, if the agent has several possible trading partners, other strategies may be superior. For instance, as pointed out in Chapter 5 the agent can try to maximize short-term profits by raising prices (and sacrificing volume) or, conversely, it may lower prices and profits per unit, in order to increase volume and market share. Thus, by changing L, the agent may try to stabilize the inherently non-linear system.

The point is that gains from production and trade subject to the AAL rule must be sufficient to compensate for excess consumption and other losses. In the Walrasian equilibrium state supply equals demand at constant market prices for all products and sectors and the welfare functions of individual agents reach a maximum and never decrease. In any complete q-cycle, which comprises production, trade and intermediate consumption, the stocks of goods and money are also constant, on average. Real economies may come close to satisfying this condition, insofar as trade flows typically exceed the short-term changes in total stocks by a large factor.

It is fairly trivial to show that the traditional Walrasian supply-demand balance condition is self-stabilizing with regard to price fluctuations: higher prices lead to reduced demand and conversely. However there is no mechanism in the system to stabilize it with respect to fluctuations in relative wealth among the agents. To investigate the stability of the system with respect to such fluctuations we solved the dynamic equations (below) numerically, based on a variety of simplified versions of the economic system. The key result of this simulation work, described below, is that the stable region of phase-space is rather small. In fact, the Walrasian system (with or without excess consumption and other losses) tends to move away rather quickly from the stable region. The destabilizing tendency appears as increasing inequity among the agents over time. Quite literally, the rich get richer and the poor get poorer. In general, it appears that long-term stability requires an exogenous regulator (i.e. government) that limits inequity, either by mandatory redistribution (via taxes) or by some other means.

6.2 DYNAMIC EQUATIONS FOR AN ECONOMIC SYSTEM

In Chapter 5 the decision rules for trade and production were established. For both adaptive agents interacting via pair-wise exchanges and also for the common price case (auction-type market) the equations defining the stock changes of the individual agents are shown as follows. Since these equations reflect the interactions of the agents, changing over time, they are effectively the dynamic equations for an economic system. We summarize below the parameters and variables previously defined, as needed to characterize the agents and their interactions:

n_a number of agents (a variable)

n_g number of goods (also a variable)

Z_α Z-function of agent α, $\alpha = 1, \dots n_a$

$X_{\alpha,i}$ i^{th} stock of αth agent, $\alpha = 1, \dots n_a$, $i = 1, \dots n_g$

M_α money stock of αth agent, $\alpha = 1, \dots n_a$,

T_α^j technology matrix of αth agent, $\alpha = 1, \dots n_a$,

$n_{\alpha,k}$ the number of different technologies available for agent α

$C_{\alpha,i}$ consumption vector of αth agent, $\alpha = 1, \ldots n_a, i = 1, \ldots n_g$
$L_{\alpha\beta,ik}$ quantity exchange coupling factor $\alpha,\beta = 1, \ldots n_a, i,k = 1, \ldots, n_g$
$L_{\alpha,j}^p$ quantity production factor for α^{th} agent, $\alpha = 1, \ldots n_a, j = 1, \ldots n_{\alpha,k}$

The theory of trade and production decision mechanisms has been described in earlier chapters. Stocks of goods and money held by individual agents will change in the course of all economic processes. However totals for the aggregation of all agents are either unchanged or are governed by exogenous processes. In the case of materials, aggregate extraction rates from the environment and depreciation rates for durables, as well as consumption for subsistence, must be considered. (To ensure mass conservation an artificial $n_a + 1$ agent 'nature' is introduced. This agent has stocks (resources), but nature does not obey the AAL rule.) A complete description – not attempted here – must also specify the exhaustion of finite natural resource stocks and the disposal of wastes. In the case of money, it is necessary to specify aggregate money supply growth, interest rates, and (if applicable) redistribution via taxes. At first we consider how a group of interacting agents approaches an equilibrium, over time. In effect, the next few sections model the operation of the 'invisible hand'.

Summarizing previous development, we can now write a dynamic equation for aggregate stocks of goods, in discrete time, assuming pair-wise exchanges

$$\Delta X_{\alpha,i}(t) = \sum_{\beta} L_{\alpha\beta,i}(v_{\alpha,i}(t) - p_{\alpha\beta,i}(t))$$
$$+ \sum_{j}^{n_{\alpha,k}} L_{\alpha,j}^p \left(\sum_{k} T_{\alpha,k}^j v_{\alpha,k}(t) \right) T_{\alpha,i}^j - C_{\alpha,i}(t) \tag{6.1}$$

Here $p_{\alpha\beta,i}$ is the price for the exchange of the ith good between agents α and β, as follows:

$$p_{\alpha\beta,i} = \frac{L_{\alpha,i}v_{\alpha,i} + L_{\beta,i}v_{\beta,i}}{L_{\alpha,i} + L_{\beta,i}} \tag{6.2}$$

The corresponding dynamic equation for the stock of money is

$$\Delta M_{\alpha}(t) = -\sum_{\beta,i} L_{\alpha\beta,i}(v_{\alpha,i} - p_{\alpha\beta,i})p_{\alpha\beta,i} + \sum_{\beta} I_{\alpha\beta} \tag{6.3}$$

where $I_{\alpha\beta}$ is the non-trade money flow (if any) between agents α and β.

These equation are non-linear, whence analytical solutions are not feasible. We have therefore calculated numerical solutions to illustrate that process.

For this purpose it is necessary to specify initial stocks, welfare (*Z*) functions and reaction parameters plus a description of the learning and adaptation process (feedback loops for adjustment of reaction parameters). Technological innovations and monetary policy of the economic system may also be specified, depending on the scenario being investigated. Finally, exogenous factors such as consumption not connected with production, taxation, depreciation, and natural constraints, if any, must also be specified.

In the analysis summarized below we have incorporated the possibility of excess (beyond subsistence) consumption and depreciation, to investigate the stability properties of the Walrasian equilibrium. Equilibrium is ensured by assuming that all gains from economic processes (trade and production) are consumed in each q-cycle.

Our minimum sectoral model of an economic system has three economic agents, corresponding roughly to sectors, namely: agriculture, industry, households. To reflect the need to compensate for excess consumption and depreciation losses we have specified minimum subsistence production levels for each agent. For agriculture, this could be interpreted as the output needed to provide food, feed and seed for the farmers and their livestock. For the industry sector, it is the requirement to maintain capital stock and compensate for wear and tear (depreciation). For households, there is a minimum food and tool requirement, to produce labor. These minimum requirements would normally be expressed as vectors that must be subtracted from the final stocks for each agent after each q-cycle.

6.3   A SIMPLE MODEL ECONOMY

The minimum economy consists of three agents, with three goods and money.

Number of agents: $n_a = 3$
Number of goods: $n_g = 3$

We selected the logarithmic form for the Z-function of agent α:

$$Z_\alpha = \sum_i X_{\alpha,i} \left(\ln\left(\frac{c_\alpha M_\alpha}{X_{\alpha,i}} \right) - 1 \right) \tag{6.4}$$

where the values of constants c_α are: Agriculture, 0.025; Industry, 0.025; Households, 0.025. The results of numerical calculations that follow show that the global properties of solutions are not too sensitive to the form of the Z-function, although the minor details ('fine structure') are affected.

The marginal subjective values are calculated from the Z-function, as indicated in Chapter 4, viz. for goods:

$$v_{\alpha,i} = \frac{1}{w_{M,\alpha}} \ln\left(\frac{a_\alpha M_\alpha}{X_{\alpha,i}} \right) \tag{6.5}$$

and the subjective value of money is given by

$$w_{M,\alpha} = \frac{M_\alpha}{\displaystyle\sum_i X_{\alpha,i}} \tag{6.6}$$

The production processes are summarized in the technology matrix (see *Table 6.1*).

Table 6.1   Technology matrix

	Food	Tools	Labor
Agriculture	1	−0.14	−0.08
Industry	−0.40	1	−0.36
Households	−1.83	−1	1

The initial consumption vector of each agent is specified in *Table 6.2*.

Table 6.2   Consumption vectors

	Food	Tools	Labor
Agriculture	0.25	0	0
Industry	0	0.06	0
Households	0	0	0.04

The fact that households must consume food (and some other goods) to survive means that production can never stop. Other economic agents, too, must consume intermediate goods in order to produce. Farmers need seeds and tools, and so on. This, in turn, implies that for the survival of the agents – even in the short term – the economic system must function. Agents can consume their stocks for a while, but if there is no trade and production, that is, if the system collapses, the stocks of goods disappear. The agent (if it is a firm) goes bankrupt and disappears or (if a worker/consumer) he/she starves. For this reason, stability is crucial.

The quantitative trade coupling factor is specified as

$$L_{\alpha\beta,ik} = 1, \text{ if } i = k, L_{\alpha\beta,ik} = 0 \quad \text{otherwise.}$$

The production coupling factors $L_{\alpha,j}^p$ are specified in *Table 6.3* below.

Table 6.3   Production coupling

Agriculture	0.35
Industry	0.29
Households	0.35

6.4 ORDER OUT OF CHAOS – EMERGENCE OF EQUILIBRIUM

In the first test, the initial state is a random selection of initial stocks, not in equilibrium. We characterize the system in terms of gross production (GP) in quantity terms. After a reasonable time (500 to 600 q-cycles) the simple model eventually 'finds' the equilibrium solution, as shown in *Figure 6.1*. This can be interpreted as the operation of the Adam Smith's 'invisible hand' or possibly Walras's '*tatônnement*'.

In a second test with another non-equilibrium initial selection of stocks, the model becomes unstable, as shown in Section 6.6. Evidently the stabilization feedbacks do not function, or are not sufficient, in all circumstances. There are two types of potential instabilities present in the model. The first arises from 'over-reaction' by the agents. This is connected with the form of the Z-function and the initial stocks. The second possible instability arises from competition between the agents, such that wealth difference between the agents increases to the point where one agent is extinguished altogether. Yet if this should occur, the model economy collapses also. Something of the sort seems to hold true in real economies, too, although the real system is far more complex.

6.5 STABILITY OF EQUILIBRIUM, SHORT TERM

If the computer calculations were perfectly accurate, an equilibrium system – such as the one described below – would reconstruct itself from q-cycle to q-

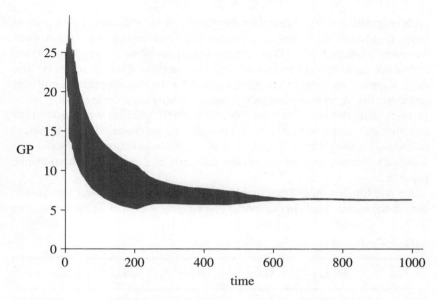

Figure 6.1   Gross Production (GP) approach to equilibrium, $w_M = 0.056$

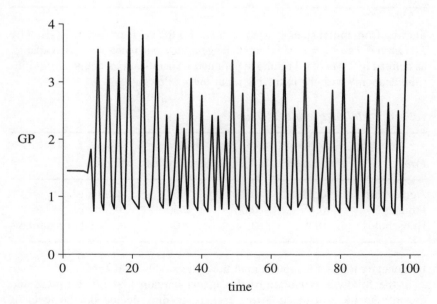

Figure 6.2   GP as a function of time, $w_M = 0.077$, short term

cycle, regardless of the stability (or instability) of the solution. However computer simulations have several characteristic features that distinguish them from real world behavior. They are necessarily simplified, with fewer agents and fewer variables and less variability, and they are finite in accuracy. The finite accuracy, to some extent, compensates for the oversimplification of real economic life. A random element – 'noise' – is introduced during each q-cycle. In noisy circumstances only an inherently stable solution reproduces itself indefinitely. On the other hand, an unstable one collapses because it cannot indefinitely satisfy the AAL rule and the self-consumption conditions. The instability mechanisms are essentially the same as described in the previous paragraph.

An equilibrium model can be characterized by the following conditions. We set the initial stocks in our model economy as shown in *Table 6.4* below.

Table 6.4   Initial stocks

	Money	Food	Tools	Labor	c_α constant
Agriculture	1000	22.98	18.51	14.07	0.025
Industry	1000	22.05	19.34	14.16	0.025
Households	1000	21.97	18.85	14.73	0.025

Starting from initial stocks, listed in *Table 6.4* the value of money is given by *Equation 6.9* as $w_M = 0.056$. For each agent only one technology is available, as defined in *Table 6.5*. The above parametric specifications define an equilibrium economy with the following input-output flows:

Table 6.5   Initial quantity I–O table

From	To Sector 1 Agriculture	Sector 2 Industry	Sector 3 Households	Total output
Agriculture	0.25	0.20	0.55	1.00 bushels
Industry	0.14	0.06	0.30	0.50 tools
Households	0.08	0.18	0.04	0.30 labor man-days

In monetary terms, the input-output flows are as shown in *Table 6.6*.

In the following simulations the agents bargain first for the price and quantity to be exchanged. After each trade they decide on the level of production. These two actions are called the q-q-cycle. Time is measured as

Table 6.6 Initial monetary I–O table

From	To	Sector 1 Agriculture $	Sector 2 Industry $	Sector 3 Households $	Total $ output
Agriculture		0.50	0.40	1.10	2.00
Industry		0.70	0.30	1.50	2.50
Households		0.80	1.80	0.40	3.00
Total $ input		2.00	2.50	3.00	7.50

the number of completed q-q-cycles. After the first time period the system returns to its initial state, as expected. In short, the model economy is indeed in equilibrium. Total production equals total consumption. In an equilibrium economy, as we have noted, stocks do not change. The gains from production of goods and labor are consumed within the sectors, as required.

As noted earlier, the equilibrium is not stable with respect to stock fluctuations. Depending on the initial choice of the stocks the simulations show that the system departs from the initial equilibrium after a relatively small number of q-cycles. *Figures 6.2* and *6.3* illustrate two possibilities, one of which

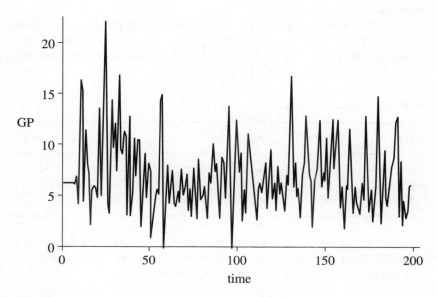

Figure 6.3 GP as a function of time, $w_M = 0.033$, short term

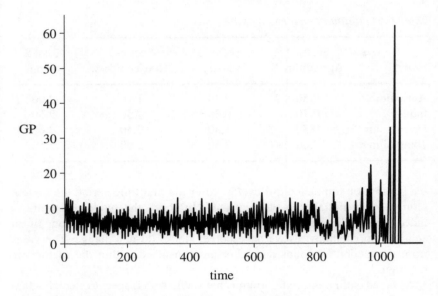

Figure 6.4   GP as a function of time, $w_M = 0.033$ long term

resembles a cyclic fluctuation around the equilibrium state while the other
resembles a more chaotic system.

  Figure 6.2 represents the state of the economy starting from initial stocks,
listed in *Table 6.7*. The initial subjective value of money is given by *Equation
6.9* as $w_M = 0.077$. Initially the GP results agree with the stable case, but after
only five q-cycles quasi-periodic oscillations appear.

Table 6.7   Initial stock vectors; $w_M = 0.077$

Agent	Money	X_1	X_2	X_3	c_α constant
Agriculture	1000	34.21	25.37	17.35	0.038
Industry	1000	32.39	27.00	17.53	0.038
Households	1000	32.26	26.11	18.55	0.039

  The effect is somewhat exaggerated in the case of initial stocks shown in
Table 6.8, when the value of money is 0.033, as calculated by *Equation 6.9*.
The change of GP in this case is more chaotic (*Figure 6.3*).

Table 6.8   Initial stock vectors; $w_M = 0.033$

Agent	Money	X_1	X_2	X_3	c_α constant
Agriculture	1000	10.69	9.57	8.31	0.013
Industry	1000	10.46	9.78	8.33	0.013
Households	1000	10.43	9.64	8.50	0.013

6.6 LONG TERM INSTABILITY

As has been emphasized repeatedly, the AAL rule is a necessary but not sufficient criterion for the long-term survival of economic agents. In our simulations the AAL rule ensures that the agents have gains in trade and production during each q-cycle. But long-term survival also requires that the gains be large enough to compensate for self-consumption, depreciation, taxes and other losses. So it can happen that, due to fluctuations in stocks, the agent is unable to compensate for the inevitable depreciation losses and self-consumption.

Our simulations show that this also leads to instability of the model economic system. Depending on the choices of initial stocks one of the agents

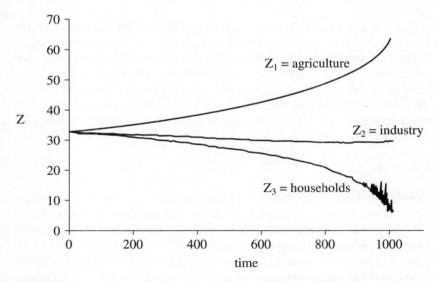

Figure 6.5   Wealth of sectors as a function of time, $w_M = 0.033$, without social tax

will become richer and richer, while another one will become poorer. This continues until the poor agent's stocks decrease below a critical level, below the self-consumption (subsistence) level. For firms this corresponds to bankruptcy level.

6.7   SOCIAL RE-STABILIZATION

The need for occasional wealth redistribution to stabilize society was realized long ago. The following quotation can be found in the Book of Leviticus (Lev. 25:10) where the laws of Moses are spelled out (allowing for some translation uncertainties).

> 10: And ye shall hallow the fiftieth year, and proclaim liberty throughout the land unto all the inhabitants thereof; it shall be a jubilee unto you; and ye shall return every man unto his possession, and ye shall return every man unto his family.
> 23: And the land shall not be sold in perpetuity; for the land is Mine; for ye are strangers and settlers with Me.

This Mosaic law – amounting to land redistribution every 50 years – was introduced to compensate for the perceived tendency for welfare (wealth) differences to increase over time. The wealthiest Jews were expected to donate a part of their money to the poor. We can simulate the results of such a rule.

Our stabilization rule is to tax 10% of the wealth difference between the richest agent and the poorest, and re-allocate it to the poorest after each cycle. *Figures 6.5* and *6.6* show the effect on the Z-functions of the three sectors for the case $w_M = 0.033$ with a social tax and without such a tax. The system now remains in the vicinity of equilibrium state indefinitely. The system is not absolutely stable; there are fluctuations, but they remain limited and apparently the lifetime of the system becomes infinite in this case as shown in *Figure 6.7*.

6.8   GROWTH

Growth means that the stocks of goods and money, and so the welfare function, are increasing. *Equation 6.1* defines the necessary economic conditions for growth in an economy with fixed technology (fixed menu of available goods and production technologies) and constant population. The 'driver' of economic growth is the desire (on the part of *H. Economicus*) for increasing wealth and welfare. To get a growth solution we must assume that all agents have unsaturated Z-functions. It is necessary, also, that the 'driving force' for trading activity and for production remains perpetually finite.

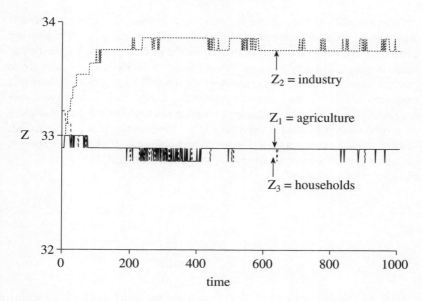

Figure 6.6   Wealth as a function of time, $w_M = 0.033$, with social tax, stable case

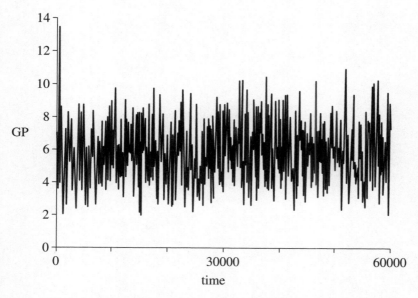

Figure 6.7   GP as a function of time, $w_M = 0.033$, with social tax

It is quite evident that, absent technological change, the above conditions cannot not be satisfied in textbook economies. First of all, in an ever growing economy, the assumption used in the derivation of *Equation 6.1*, namely that nature is very large compared to the economic activity, so it is an infinite reservoir, does not hold. The scarcity of inputs will stop growth. Further, if there is no increase in the menu of possible goods, and their quality (perform-ance) while continuously reducing the material requirements for producing them, economic growth must inevitably slow down and stop as Z is saturated. We consider the origins and processes of technological change in Chapter 8, after some preliminary discussion of aggregation problems.

7. From agent to aggregation

7.1 THE TRANSITION FROM MICRO TO MACRO

Up to now we have focused mostly on so-called 'micro-foundations' and implications for the behavior of individuals and small groups of agents (belonging to *H. Economicus*) interacting within a relatively stable environment, not too far from equilibrium. However, in order to address macroeconomic problems, including such phenomena as 'bubbles', depressions and growth, there comes a point where it is necessary to make the transition from micro- to macro-perspectives. This transition involves a major change of perspective: a number of variables and factors that are 'exogenous' at the micro-scale become 'endogenous' at the macro-scale. However, first we need to deal with some preliminaries, especially definitions of terms.

An economic system consists of a collection of economic agents, consisting of firms or households, performing economic functions, including exchange, production or consumption for subsistence purposes. (Consumption for other purposes plays an important role in creating demand, for instance, but as noted in Appendix B it does not always satisfy the AAL rule that governs most economic activities.) Agents necessarily interact within an institutional (and a natural) environment. In economic processes the agents exchange goods, money and information with each other, as well as exchanging materials and energy with the natural environment.

The real economic system is inherently non-linear, with many feedbacks, some of which are inherently stabilizing (i.e. positive feedbacks), while others are destabilizing (negative feedbacks). These feedbacks are mutually interdependent. The economic system is extremely complex in the technical sense of the word. It has *memory*: economic behavior is generally history dependent, hence it is *path-dependent*.[1] In the real world, of course, some of the destabilizing processes are non-economic, including civil conflicts and wars. We neglect these, for the most part. Macroeconomic processes with negative feedback include hyperinflation, deflation, and consumption of capital – such as seed for next year's crops – for immediate survival. More subtle examples include 'bubbles' in which asset values are blown up by exaggerated expectations – usually fed by the investment frenzy itself. This is invariably followed by a collapse, triggered by a spreading realization that 'the emperor has no clothes', that is, that the assets supposedly underlying the previous expansion were imaginary.[2] Each bubble begins with positive feedbacks – growth feeds expectations of further growth – and switches suddenly to a catastrophically negative feedback during which losses encourage behavior that impedes growth.

The primary protection against losses from such events, both at the societal level and the individual level, is learning how to avoid them. When negative feedback situations do nevertheless occur, from time to time, it is the task of macroeconomic policy-makers to find ways to counteract them. After the first phase of the Great Crash of 1929 the US Federal Reserve made the looming problem of wealth destruction much worse by tightening credit, based on orthodox economic theory. But the fact that much of the previous runup of stock prices had been based on credit – that is, through buying on margin – meant that the central banker's actions triggered a further sell-off. As more wealth was destroyed by the fall in share prices, consumer confidence fell in parallel, spending stopped and the Great Crash of 1929–30 morphed into the Great Depression of 1930–35. John Maynard Keynes proposed the appropriate countermeasure – deficit spending by the government to kick-start demand – but it was fiercely resisted at first by orthodox conservatives, and never implemented on the necessary scale until the beginning of World War II.

However this is not the place to review past history or critique monetary or fiscal policy. The only point we wish to make here is that policy-makers in the 1920s and 1930s had not yet learned how to anticipate and avoid or counteract negative economic feedbacks. Some of the learning necessary to avoid these problems has evidently occurred since then; certainly deficit spending and tax cuts to boost spending are now part of the arsenal of macroeconomic tools, along with interest rate manipulation. Whether the level of learning at the appropriate levels of government is now sufficient, is still debatable (and hotly debated), of course. But we need not contribute further to that debate here.

At the micro-level an agent's wealth may increase or it may decrease, according to the ups and downs of macroeconomic circumstances and policies: It is a political article of faith that 'a rising tide lifts all boats'. The converse is no less valid. Avoidance of losses by individuals and firms is only possible up to a point. Unavoidable losses of wealth from such large-scale events are among the problems that may confront individual economic agents.

During the course of action and interaction all economic agents change and evolve. The institutional and natural environment also changes, as a result. It co-evolves. We distinguish two kinds of change mechanisms operating in an economic system at the systems level. The first of these is short-term cybernetic regulation by information feedback, based on signals (such as aggregate commodity stock and flow changes and price changes), together with fiscal or monetary policy intervention at various governmental levels.

The second mechanism is long term. It involves learning-by-doing, learning-by-using, adaptation to changing circumstances by trial and error, invention and innovation. The latter two responses, leading to new products and new sectors, require more detailed discussion (in the next chapter). Learning-based changes begin with modifications of expectations. They are reflected in the dynamic parameters of the L-tensor, and (to some extent) the input-output coefficients T_j, reflecting incremental evolutionary changes arising from interactions with the changing institutional, socio-political and natural environment.

7.2 FROM INDIVIDUAL WORKERS SELLING LABOR SERVICES TO LABOR SUPPLY

Labor services are clearly one 'factor of production' at any scale. Labor inputs measured in worker-hours are relatively easy to aggregate, and this is done routinely (for non-farm labor) by the US Bureau of Labor Statistics (BLS), and comparable agencies in other countries. These agencies also calculate average wages (by sector) from total wages and average hours worked. The BLS estimates unemployment at any given time by subtracting actual hours worked from 'potential' hours worked, based on estimates of the number of workers seeking jobs at any given time and registered for unemployment benefits. Official employment statistics are based on two different surveys, one based on data from employers and the other from households.

Of course the labor force in a modern economy is very heterogeneous, due to differences in age, region, gender, race and educational status. Simplistic aggregation on the basis of hours worked begs the important question of differences in quality of labor (or *human capital*, as it is sometimes called). One possible response is to measure value in terms of wages or earnings. This would mean measuring the total labor input into the economy in terms of total wages and salaries. But since wages and salaries constitute the greater part of GDP, this would effectively confuse outputs with inputs, making any measure of productivity impossible. (This has long been a fundamental difficulty in measuring the productivity of services.)

Other approaches to measuring labor quality have been tried, notably in terms of education and training inputs. But years of formal schooling are very imperfect measures of labor quality. In the first place, other forms of job training and experience are neglected, even though they may be more important. In the second place, even for jobs with comparable formal educational requirements, measured in years, there are enormous differences in remuneration, depending on other factors.

The problems of defining and measuring aggregate labor are, in fact, very difficult and not to be dismissed lightly. However, the difficulties are fairly well understood, and after all is said and done, most economic models still utilize the total of hours worked as a primary variable (factor of production). We can add nothing useful to the discussion.

Before passing on to other matters, it is important to acknowledge that a new conceptualization, *human capital*, has found its way into the economics literature since the mid 1980s. The general idea is that human capital is an accumulation of knowledge and skills, embodied in human beings (as well as books, computer programs, designs and pictures) and that it is this knowledge base, rather than the material goods in which it is (partially) expressed, that really creates wealth. The growth of human capital resembles the growth of wealth in our sense, and it differs from the older conceptualization of capital in one critical respect: human capital (because it is really knowledge) *does not depreciate*. Indeed, in the so-called endogenous growth theory, where human capital replaces either conventional capital or labor (or both) the old familiar constraint of constant returns to scale can be discarded: It is argued that human capital allows for *positive returns*, resulting from so-called 'spillovers'.[3] Indeed, it is positive returns from spillovers that is supposed – in the endogenous theory – to account for economic growth. However, the major (and, in our view, conclusive) reason for not introducing human capital hereafter as a factor of growth is that it has never been adequately defined, still less measured.

Measurement of human capital is a major problem in the context of estimating national wealth. A plausible approach is to focus on expected future earnings (returns) by workers. These earnings can – in turn – be estimated from expected economic growth, together with some hypothesis about the allocation of national income between labor and capital. In other words, the value of human capital today depends on the future prospects of the economy. But that, in turn, depends upon the investment in human capital … and so on. In short, the dependence is recursive and bi-directional. Hence the measurement problem is unlikely to have a straightforward solution.

7.3　FROM PRODUCER GOODS OWNED BY AGENTS TO AGGREGATE CAPITAL

The term capital, in its modern sense, has evolved from its original meaning as land, farm implements, tools and domestic animals. In the early nineteenth century it was associated with a political philosophy called *capitalism* which, while never precisely defined, became one pole of a long-running and continuing conflict with its polar opposite, *socialism*. The difference between these two philosophies, as seen in retrospect, seems to be differing views on the role of profit and the need for investment. The term *capital* carries a heavy load of ideological associations, some of which are irrelevant to this book.

In our theoretical development up to this point, capital consists of goods and money belonging to economic agents (e.g. firms) that are producers. Goods in stock are part of a firm's capital, whether or not it is ultimately embodied in products. Similarly, a firm's money is part of its capital, whatever its immediate use. The distinction between *material* capital, whether durable or not, and *monetary* capital is not important for our purposes.

At first glance, it seems that there is another type of capital, not associated with firms, namely public-sector capital or 'infrastructure'. It is perfectly clear that the productivity of the capital owned and controlled by agents or firms is also affected by the condition of the roads, bridges, harbors, water and sewers, electric power supply and other utilities. However, while most infrastructure in many countries is produced and maintained by the public sector, this need not necessarily be the case. The primary reason for public ownership, in most countries, is that infrastructure projects tend to require large-scale confiscation of land and consequent interference with private activities, thus necessitating adjudication for purposes of compensation, and enforcement where necessary. The scale of capital expenditures involved often exceeds the capability of the private sector, as well. But once the infrastructure is in place, there is no reason it cannot be partly or wholly privatized, as is happening in many countries at present. The major problem in doing this is that the end result may be to create monopolies. However this subject is beyond our scope, here.

In the spectrum of economic agents, there is a group that combines two roles. One role is that of *producer*, mainly of labor, but also of a variety of 'home-made' goods, for domestic consumption, barter, or sale. The other role is that of *consumer*, not only of goods, but also of services. (Given that the ultimate output of the economy is services, it is tempting to regard all durable goods as a form of capital, but this is not our point of view.) The present situation in industrialized countries is such that most of the production and consumption of goods that occurs within farms, for instance, has been monetized. Only domestic work, especially by women, remains largely non-monetized.[4]

Prior to the industrial revolution in Europe most food, clothing and household goods were home-made by the vast majority of workers/consumers. They were not bought or sold to a significant extent in markets. The same was true of human labor: in the middle ages peasants traditionally worked for the landowner in exchange for the use of a fraction of the crop, or a small plot of land and a crude house for personal use. Peasants (i.e. most people) were effectively outside the economic system. The landlord – as likely to be a monastery or bishop as a feudal lord – took most of the saleable crops, including wool from herds of sheep. 'Surplus' labor was largely consumed by hierarchical superiors for building unproductive structures such as castles, churches or cathedrals.

The Renaissance and the Protestant Reformation changed the old regime and opened the door to trade and commercial activity. Most of the new wealth – both goods and money, as well as houses – remained in the towns, of course. However even the farmers accumulated stocks of tangible goods, including home-made furniture, implements, clothing and blankets, pots and pans, as well as houses and barns. These items were passed from generation to generation through inheritance. In due course they, too, gradually entered the market economy and acquired monetary value.

Inasmuch as durable goods provide services to consumers, it is tempting to equate such accumulations of durable 'dual purpose' goods in residences with capital. Yet this is troublesome. Tools and implements are clearly productive, as are watermills, plow horses, cattle, pigs, sheep and arable land itself. But most rural families shared quarters with their animals. Should the house be considered productive or not? What of the road and the bridge over the stream? Workers had to eat in order to work. Should the kitchen be considered part of the productive capital? What of the clothes of the workers? The distinction between *production* and *consumption* might seem to be very clear today, or even for a wealthy *rentier* or landowner in the past, but hardly for a farmer, or his wife.

It is hardly surprising that Marx and other nineteenth century radical reformers thought that the distinction between 'working class' or 'proletariat' – people who worked for a living – and property-owning class (or 'capitalists') who consumed the products of other people's labor, was so clear and natural. But the distinction has eroded, in part because more and more people are both workers and property owners, and because the concepts of 'goods', and 'production' and 'consumption' have become blurred. The word 'consumption' as applied to a house or a car does not convey the same idea as when applied to food or drink or soap or fuel. Houses and household goods are not really consumed. They may be broken or worn out, or even obsolete, but they do not get 'used up'. Hence the interpretation of consumer durable goods as producers of immaterial services seems preferable to us, even though it does not correspond to current accounting practice. This leads us back to the problem of definition of capital.

The idea of capital as a 'factor of production' is quite an old one. The term 'capitalism' presupposes it, of course. Marx focused attention on the concept intensively with his critique (Marx 1867).[5] While Marx's theory of class warfare has not survived, the notion that surplus income reinvested in various ways contributes largely to future productivity has become a core concept of economics. A key feature of the single-sector neoclassical models of the 1950s is the explicit introduction of aggregate production functions in which capital services are derived from an artifact called 'capital stock'

Capital stock, more or less abstract, was deemed to be one of the key variables explaining production and growth during the late nineteenth and early twentieth centuries. There was much attention to capital-labor and capital-output ratios, for instance. However, in physical terms, capital is unmanageably heterogeneous. For this reason, many economists have questioned whether there is any consistent way of measuring capital stock independently of the income it produces (its 'rate of return'). It is the returns to capital in the form of interest, dividends, rents and royalties that are measured in the national accounts. The implication is that one might work back and measure capital stock as the (hypothetical) quantity that produces the measured return.

This notion makes sense if (and only if) the rate of return is somehow fixed exogenously, perhaps equated to an unchanging discount rate or a rate of time preference. In the real world rates of return on explicit financial investments can vary enormously, even from one day to the next, as current valuations change in response to changes in expectations of future returns and perceptions of risk. In other words, actual returns depend on expectations of future returns, which can go up or down from one day to the next. These questions gave rise to the so-called Cambridge Capital Controversy (CCC) between two groups of economists, one associated mainly with Cambridge, UK and the other associated mainly with MIT, in Cambridge, Massachusetts. On the UK side the instigator and leader was Joan Robinson (Robinson 1955b, 1955a). On the US side the best known figures were Robert Solow and Paul Samuelson. The CCC was later reviewed by Harcourt (Harcourt 1972). We need not recapitulate the arguments, but a useful summary can be found in Mirowski (Mirowski 1989a, pp. 338–50).

Nowadays, for better or worse, total fixed capital stock is usually calculated in monetary terms by the so-called perpetual inventory method or PIM, which starts from a base year and adds new investments in various categories (e.g. residential housing, nonresidential buildings, machinery, roads and bridges, etc.) at current prices adjusted to a standard year, while simultaneously depreciating existing capital stocks based on assumed lifetimes (Maddison 1995).

Evidently different types of capital (e.g. housing, infrastructure, machinery, vehicles, etc.) as mentioned in the previous paragraph differ in functional ways. And they differ in terms of lifetimes (depreciation rates, obsolescence rates) and in terms of productivity. Finally, they differ in terms of flexibility or substitutability with respect to each other and with respect to other factors of production, notably labor and energy (or exergy). Here the important point is that some types of capital are productive without other factors, some can effectively replace other factors – as some machines can replace some kinds of labor – and some are only productive as long as other factors are available also. For example, machines normally require fuel or electric power whereas housing and some infrastructure may be productive themselves.

The mix of long-lived capital (infrastructure and structures) and short-lived capital (e.g. vehicles and computers) has shifted significantly in favor of the latter, over recent years. The rate of depreciation was much lower (of the order of 3% p.a.) a century ago than it is today. The current rate of depreciation of fixed capital is probably at least double what it was in 1900 and could be even higher. There is, unfortunately, no direct way to measure depreciation rates. As a consequence, aggregate capital stock is an artifact, based on money flows. Capital stock calculations are made for the most developed economies by national institutions, notably the Commerce Department in the US and the national statistical offices of other countries, as well as by the OECD.

In the future it may be useful to distinguish 'normal' capital from capital dedicated to information/communications technology (ICT), in particular. The primary argument for this is that ICT capital – apart from fixed telephone lines – depreciates much faster than buildings, machines and other equipment. The fraction of total capital that is ICT related has been estimated by Jorgenson and Stiroh for the US (Jorgenson and Stiroh 1995, 2000) and by Takase and Murota for Japan (Takase and Murota 2004).

Apart from the fluctuations in the monetary valuation of different forms of capital, it is important to bear in mind that some capital investments yield non-monetary returns. This applies, of course, to owner-occupied housing and private cars, as well as other personal property. Since the history of economics is, in some sense, a history of the monetization of goods and services, it is reasonable to adopt the view that the process of monetization will continue. Even now, US tax authorities are increasingly insisting on calculations of 'rental equivalent' for house property, for purposes of assessing income tax liability. Personal property (mainly automobiles) is also taxed in some jurisdictions. In effect, some consumer durables are already being considered (for tax purposes) as if they were sources of income.

7.4   MONEY AND MONETIZATION

Money has been treated as a given for economic agents up to this point. That means that economic agents are assumed to utilize any or all of the standard forms of money – namely, currency, travelers' checks, checking accounts, savings accounts and money market funds – but that, as economic agents they do not, and cannot, affect interest rates, conditions for borrowing, or rates of inflation.

Money is perhaps the easiest variable to measure (by counting) and aggregate. This is routinely done by the US Federal Reserve Bank and other central banks. The usual classification of money into M1, M2 and M3 reflects decreasing availability to settle transactions. M1 consists of immediately available funds, M2 includes retail savings accounts, retail money market accounts and short-time bank deposits, M3 includes large savings accounts, wholesale money market funds and long-time deposits. Securities, such as bonds and treasury bills (T-bills), are not included although of course they can be sold or used for collateral, as can shares. Debit cards are simply a substitute for checks or cash.

Credit card debt is also not included, although credit cards are widely used for transactional payments. The reason is that the bank makes the payment immediately and the customer repays the bank after a short delay, when the bill is received. To count credit card debt as new money would actually be double-counting. However there is no doubt that unsecured credit cards have greatly increased the ability of consumers to spend before they have earned.

But, at the same time, in an economic system considered as an entity, there are other kinds of money equivalents and, indeed, some conceptual problems that deserve serious attention. There is a very large technical literature on the subject, but it is comparatively obscure. The problem of defining money, mainly from an historical perspective, is discussed in Appendix A.

Monetization of the economic system has occurred gradually, as already mentioned in connection with domestic production and consumption. The process was accelerated in Europe in the middle ages as non-ecclesiastical landlords needed money for taxes (to pay for the dynastic wars of kings) and for luxury goods – imported from the East – and weapons (notably cannon) that could not be made at home, so to speak. Landlords increasingly demanded cash rents from their tenants, who began to sell produce in the weekly (later daily) markets in towns. Rural wealth increased, to some extent, automatically as domestic animals reproduced themselves and crop plants produced seed. Farmers constructed fences to keep sheep and cattle from straying and sheds to house pigs and chickens and protect them from cold and predation. Tenant farmers with surplus money, in turn, purchased items in the market they could not make themselves, from shoes to plowshares. As more money circulated, other goods – ceramics ('china'), glassware, silverware, luxury fabrics, tobacco and spices, even books, clocks and musical instruments – appeared in the market and new trades created or imported them.

In effect, the availability of money and credit enabled and accelerated trade, which subsequently created new opportunities. Economic growth since the middle ages has been driven, to a significant degree, by the process of monetization.

7.5   RESOURCE INPUTS

Resource inputs of a given kind (e.g. energy carriers, biomass, metals and so forth) and goods produced, of a given kind, can also be added up, and aggregated, once the appropriate unit of measurement has been fixed. Because material stocks and flows are so essential to our world view, it is also useful to consider certain aggregate material (commodity) flows explicitly. Economic models normally reduce everything to monetary value terms, and that is one possibility. However other choices of units are also possible. Mass is one possible choice of unit, although it has almost no economic significance. Adding masses of different materials serves no useful purpose. In the case of fuels one can easily cumulate in terms of heat-of-combustion – or *available energy* – content.[6] This is quite commonplace in government statistics or economic studies of resource issues.

A further argument for caution in aggregation arises from the physical nature of extractive materials. At least four major groups suggest themselves. One group consists of so-called *renewables*, especially biomass (agriculture, forest products, fisheries, etc.) A second group consists of fossil fuels. A third category consists of other *exhaustible* mineral resources, including topsoil. The last category consists of rainfall, flowing water and air, which are provided free of charge. For purposes of environmental impact analysis the distinction between renewables and exhaustibles is important, as is the role of non-priced material inputs (i.e. water and air). The distinction between fuels and other exhaustibles is useful for purposes of climate policy studies, since most of the so-called greenhouse gases are generated by fuel combustion, whereas most of the other pollutants of importance are the result of other conversion processes (especially mining, smelting and chemicals).

7.6   SECTORS AND CHAINS

Much economic analysis, and especially growth theory, is traditionally addressed in terms of single-sector models, with an undifferentiated product. The existence of fundamentally different types of labor, materials and capital suggests that the economy cannot be realistically represented as a single sector, producing an undifferentiated product from undifferentiated labor, capital and materials. This introduces the problem of appropriate *sectorization*, which we consider in this section.

Common sense (and the previous sections) remind us that there are several fundamentally different types of labor, capital and materials (as well as money), and also fundamentally different types of products. Indeed, one of the most important differences on the output side is between goods and services. Another is between goods for final consumption and capital goods. Finally, there are fundamental differences, for example, between goods that are destined for final consumers and goods that are transformed during an intermediate production process.

In *Chapter 3* we distinguished four types of economic agents according to their role in the system. Thus Type A agents perform unit processes based on the use of human labor, but little or no capital, for collecting and transforming 'gifts of nature'; Type B agents perform unit processes, using labor and capital for transforming materials produced within the system into other products with higher value. Type C agents are merchants, who use labor and capital to buy or sell but not to transform, while type D agents are consumers that buy consumer goods and sell services, including labor (but also other services).

Pure Type A agents are scarce nowadays. The category comprises subsistence farmers and fishermen, and some artisanal miners and prospectors. Type AB agents combine extraction with processing. They also perform downstream conversion processes requiring capital equipment as well as labor. Examples of this kind include hydroelectric power plants, tree farms and plantations, modern 'industrial' farms, high-tech fishing, petroleum and gas operations, and large-scale mining or quarrying. It is also worthy of note that all industrial processes requiring water, and all processes (including final consumption) that use fuels and depend upon oxygen from the air are Type AB by this convention. Air and water are the only free gifts of nature not requiring labor or capital to extract and utilize. Nature also provides non-material inputs such as sunlight and climate, but they are not subject to mass balance conditions.

In Chapter 3 a quasi-cycle was defined for an agent performing a unit process, the key point being that after a certain number of operations the state of the system can be compared directly and objectively with the initial state, in terms of stocks of goods and money. We noted that the AAL rule applies to quasi-cycles in a very straightforward way. An agent of Type A, B or C that does not satisfy the AAL requirement consistently cannot survive.

A process chain is a sequence of linked q-cycles, beginning with the extraction of raw materials and ending with a finished product. For a product or commodity that is sold in the marketplace, value and price can be equated. This condition is most easily applied in a market economy, where values are determined by market prices. However, the condition can also be applied in a Robinson Crusoe situation, where the only criterion for value is subjective.[7]

(In effect, Robinson Crusoe spends his time on projects with the highest subjective value to him., such as catching fish or finding edible fruits or nuts, or making fire. He will not spend time on a project with a very uncertain or low value outcome, such as digging drains or sending smoke signals when there is no ship in sight.)

The AAL rule implies that the value of the final product at the end of the process-chain must exceed the sum total of values (costs) of all inputs, including labor and purchased intermediates. The surplus value can be regarded as profit, rents (payments to the owners of natural capital) and payments to produced capital services. The sum of all payments to labor and payments to capital can be equated to value added in a market economy. For the economic system as a whole, it is evident that a necessary condition for long-term *sustainability* is that value-added minus payments to labor should be at least sufficient to replace capital depreciation, including both produced capital and natural capital (Pezzey 1989; Solow 1992; Pearce and Atkinson 1993; Toman et al. 1995; Pearce et al. 1996).

Aggregation can be safely carried out within categories, or even along process chains, although that is more difficult. While one may aggregate within categories, aggregation across categories is likely to be misleading. In effect, there is a very strong argument for sectorizing the macro-economy into categorical 'cells' so as to reflect the various distinctions between agents, processes and materials categories that we have made at the microeconomic level.

As noted, pure type A agents who exploit gifts of nature by means of labor alone are rare enough in a modern economy to neglect. Type AB and Type B agents correspond to extraction and refining, manufacturing and construction, while Type C agents are traders and Type D are service providers (and consumers). The implication is that the simplest economic model with the necessary attributes would involve three sectors (extraction, conversion and service) as illustrated in Chapter 6 (Section 6.3) and again in *Figure 7.1*.

A further subdivision of the extraction and conversion sectors to distinguish fuels and energy carriers from other goods would increase the number to five (also *Figure 7.1*). A further subdivision of the conversion sector to distinguish capital goods from consumer goods would make six sectors. Adding a government sector (below) would make seven. We have neglected back-flows, for simplicity, but they obviously play a role. Unfortunately, even a three sector non-linear model is extremely difficult to solve.

As it happens, most economic models are either single sector or many sector (input output). Evidently a single sector model is too unrealistic for most purposes. Such models miss some critical behavioral features. But many-sector I-O models are perforce linearized, static, and data constrained. They can only be dynamized (within limits) by artificial manipulations, in-

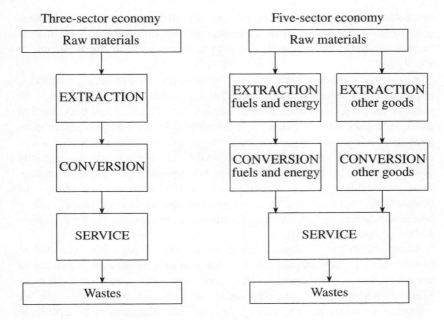

Figure 7.1   Three- and five-sector models of the economy

corporating a number of simplifying assumptions. Neither the single sector models nor the multi-sector I-O models can explain economic growth in the real world.

The special role of government is discussed next.

7.7   GOVERNMENT

As we pointed out at some length in Chapter 1, human society does not consist only of economic activities and *H. Sapiens* is not pure *H. Economicus*. Free competitive markets cannot exist for long without some degree of regulation, police and enforcement, which is one of the primary and earliest functions of government, and of *H. Custodius*. Modern societies demand other functions, including defense against external threats, economic management via monetary and fiscal policy, creation and maintenance of physical infrastructure (roads, harbors, tunnels, bridges, water supply, etc.), protection against natural disaster or epidemic (dikes, dams, weather forecasts, sewage treatment, etc.), education (schools and universities), basic science, environmental protection, protection of national cultural heritage (which includes religious institutions in some countries), and social security for those least able to help themselves. The latter category is variously defined; it may include the poor, handicapped, seriously or chronically ill, unemployed, or simply old.

All of these functions are carried out (mainly) by agencies of national, regional or local government, and paid for mostly by taxes imposed on personal income, corporate profits, consumption, value-added, trade, capital gains, inheritance or property. In some cases it is feasible to finance infrastructure by fees-for-service (e.g. highway or bridge tolls, charges for water, etc.) Some social services, such as schools and hospitals, can be partially self-supporting, by charging for service but providing discounts or exemptions for qualified groups. Countries vary considerably in their choices of functions to be supported, levels of support and method of finance, all of which are determined by a political bargaining process, not by the operation of free competitive markets.

The point is that government agencies are not, in principle, expected to make a profit. They are not (in general) financed by sales of products or services, though a few agencies may come close. The people who run these agencies, and the politicians who decide on priorities and policies, are not primarily acting as members of *H. Economicus*. Although some economic theorists, especially of the so-called Chicago school, have argued that many government activities can be understood on the basis of economic principles, we think that people who make their careers in the public sector are, at least during working hours, more likely to be members of the subspecies *H. Custodius*. Nevertheless government agencies are consumers of goods and services from the private sector, and providers of services to the private sector. Most of the services provided are not paid for directly, as such. Even where fees are charged, they are politically or administratively determined. There is no bargaining of the kind we have analyzed in earlier chapters, especially Chapter 5.

We have chosen to treat taxes as a kind of automatic charge or wealth loss to taxpayers, in much the same sense as depreciation of physical capital is a wealth loss. Survival in the competitive marketplace requires that profits from production of saleable goods and services suffice to compensate for these losses. From this perspective, government agencies are consumers of resources and producers of social services that cannot be measured directly in most cases, except in terms of their cost of production. For accounting purposes, therefore, it is convenient to regard government agencies (along with some other institutions) as not-for-profit firms.

The remainder of this chapter is devoted mainly to a theoretical analysis of the dynamics of economic systems *qua* systems.

7.8   SHORT-TERM REGULATION AND COORDINATION: MARKET PRICES

One of the most fascinating problems for economists over the past two-plus centuries has been to explain the workings of the 'invisible hand' (in Adam Smith's memorable phrase) whereby selfish individuals interacting through a free market achieve a socially beneficial outcome. The modern version of this idea has been rephrased (and oversimplified) in some conservative quarters as a slogan: 'greed is good'. There are very good reasons to doubt that the invisible hand of the market is always benign or that policy should be based on the assumption that it is. The importance of trust in economics is discussed in Appendix A in connection with the role of money. In fact, trust is important throughout the financial world and, indeed, throughout the economic as well as the political system. The nasty consequences of failures of public trust have made themselves uncomfortably felt in the past few years. However, the problems of inadequate disclosure and poor corporate governance have received a great deal of attention recently and need not be discussed at great length here. The question we need to address first is the nature of the feedback mechanisms that tend to pull an aggregation of agents interacting through a market toward (or away from) equilibrium.

Traditional neoclassical theory is vague about how this adjustment process works, since the original theory presupposes a Walrasian equilibrium between supply and demand, which assumes a semi-miraculous process (*tatônnement*) for price information dissemination and determination. More realistic alternative mechanisms, such as auctions, have been proposed, but they also do not adequately describe the mechanism (or several mechanisms) whereby market prices – and changes – actually become known to competing agents. To be sure, technology has moved on from Walras's time. Today the several stock-exchange mechanisms for buying and selling shares and other financial instruments are familiar and reasonably effective (except during panics.) In fact a mini-industry has grown up around the collection and publication of price data for stocks and shares, and for commodities and commodity futures. Nevertheless, it is important to recognize that a number of variations on the theme exist today (e.g. direct bilateral bargaining, several kinds of auctions, use of market specialists or arbitrageurs) and still others may be invented in the future. All of these mechanisms involve the use of money as a medium of exchange, of course.

Yet, there are exchange mechanisms that do not require money or explicit price-determination. The oldest of all, barter, is an example. Barter arrangements can be explicit (as in the Jack and Jill example described earlier) or implicit, as when one party does a favor for another, accumulating 'brownie points' and hoping for a return favor (unspecified) in the future. Or, in some

Eastern religions, the repayment is in a vaguely defined 'Karma' account that can be 'cashed' in (so to speak) in a later existence.

The mechanism in question has already been explained in Chapter 5 for individual agents interacting in pairs. The 'force law' for price-taking traders (*Equation 5.11*) can be extrapolated automatically to aggregates (as already noted) by simply assuming that the price p is exogenously determined by the market. A higher market price increases the difference between subjective value of the good and the price offered, thus increasing the incentive of a seller to sell and decreasing the incentive of a buyer to buy for stock.

In the case of producers (*Equation 3.22*), the mechanism is slightly more complicated, but similar. If there is latent demand and unutilized capacity in an industry, supply can adjust quickly by adding labor-hours and utility services. If capacity is limited, new capacity can be added – albeit not instantaneously – by expanding or replicating existing plants. If the price increase is applicable to the product (and not, by assumption, to the stocks of purchased raw materials or intermediates on hand) the incentive to produce from stock is increased. Conversely, if the price increase applies to purchased inputs but not to the product, the incentive to produce is decreased. The net result is that higher prices depress demand and generate increased supply, and conversely.

At the micro-level, market prices are not really explained. They are either set by a monopolist producer or a monopsonist buyer, or determined by a myopic pair-wise bargaining process that depends only upon current stocks and preferences. At the macro-level, the prices of all goods in equilibrium are assumed to be determined by the balance between aggregate supply and demand, for each good. The actual (quantitative) form of supply and demand curves at equilibrium are exogenously determined, but not explained by the standard theory. We do not offer a quantitative theory to explain the relative prices of specific goods (in terms of labor hours or wages, for instance), except to note that prices are determined by natural availability of raw materials and the state of extraction and processing technology, as well as the costs of produced inputs.

Natural availability – or scarcity – of metals, for instance, depends on the distribution in the earth's crust, which (in turn) is related to known – but not perfectly known – geological features. Thus the practical availability is really a function of discovery, which depends in part on luck, in part on geological knowledge and in part on the intensity of the search, both past and present. Gold has been sought intensively for thousands of years, whence new discoveries are rare and limited to very remote regions. Magnesium, titanium and uranium, on the other hand, were not considered resources until the mid twentieth century.

The other factor in price determination is the state of extraction and processing technology. Here the role of luck is small, and the role of general scientific knowledge is larger, while the role of specific engineering knowledge is very large indeed. The latter, in particular, is driven by market forces. The main point to be emphasized for present purposes is simply that relative – and therefore absolute – prices can change quite rapidly, either because of new discoveries (e.g. of large underground deposits of oil or gas) or because of technological breakthroughs in extraction or processing or new users.

Anticipation plays a role, and one of the functions of prices is to signal future changes in supply and trigger appropriate responses in terms of reallocation of resources, primarily capital investment. Market prices are useful as signals, because they are relatively easy to monitor and measure, whereas quantities traded are sometimes confidential. In competitive markets – where there are no monopolies or oligopolies – it is reasonable to assume that all the agents are price-takers. In this case, one can describe the behavior of a neoclassical economic system in terms of equilibrium prices.

However, this simplification is not needed. In our non-equilibrium approach prices are always the consequence of interactions between agents, with no restrictions as to relative wealth or market power. Economic agents make their decisions based on their stocks of goods and money and their expectations. The latter are based on experiences and knowledge. The marginal subjective values and the dynamic parameters (L-matrices) embody that accumulated knowledge and resulting expectations. Together, they determine the prices at which exchanges take place.

Before passing on, it must be acknowledged that competitive markets and price signals do not always lead to long-term stability. The mechanisms that normally operate to increase supply in the face of a perceived shortage do not always work. For instance, it is fairly well known that when a desirable species of bird or animal or fish becomes very scarce, the market price will rise and the intensity of the hunt for the last few individuals is quite likely to result in their extinction. This is because the rate of reproduction is fixed by nature and cannot be accelerated by demand. For example, Siberian tigers are now so scarce that one of the few surviving males may never encounter a female in heat.

A further comment on stability arises from the obvious point that successful traders or manufacturers are better able to withstand setbacks than their less successful colleagues, whether from their own occasional bad judgments or from causes beyond their control. In short, in a competitive free market, as time passes success breeds success, the rich tend to get richer – often by combining into oligopolies or monopolies – and the poor tend to be squeezed out. The inevitable result is increasingly inequitable distribution of wealth. If this tendency is not compensated by external forces, notably government regulation and/or redistributive taxation, society will eventually become so-cially unstable, as has happened more than a few times in human history. We discussed this instability problem in Chapter 6.

7.9   WEALTH MEASURES AND GDP

Moving from individual agents to aggregates, it is tempting to assume (as did the dictator at the end of Chapter 5), that the total economic wealth of an economic system can be formally written as the sum of the wealth of all the individual agents, namely

$$Z = \sum_{\alpha} Z_{\alpha} \tag{7.1}$$

where Z_{α} is the wealth of the αth agent. However this formulation neglects common property goods, like parks, air and fresh water, fisheries and bio-diversity that cannot be owned and must therefore be treated as a special category.

It also neglects interaction effects among agents, of which congestion and pollution are only the obvious examples. There is little doubt that many people prefer to live near other people, if only to reduce the costs of produc-tion, exchange and distribution – but also to facilitate positive social interactions. On the other hand when the population density becomes too great congestion begins to increase the costs of production and exchange, and there are increasingly negative social interactions (such as crime) necessitat-ing expenditures on security. Similarly, the more agents are producing and consuming in a limited area, the more serious the costs of waste treatment or disposal.

Kenneth Boulding coined the terms 'cowboy economy' and 'spaceship economy' to characterize these extremes (Boulding 1966). In the cowboy economy people are far apart, there is no congestion, and resources are effectively unlimited. In the spaceship economy resources are scarce, wastes must be recycled, even at high cost, and people get in each other's way. *Equation 7.1* does not reflect these aspects of wealth.

Furthermore, the summation in *Equation 7.1* is akin to adding oranges and apples. This is because the individual wealth functions are evaluated by the agents, in subjective (internal) units, and there is no common unit of meas-urement applicable to all agents. Intuition suggests, however, that the total wealth of the country – disregarding the class of problems mentioned above – may be reasonably approximated by introducing a universal value of money. Taking this approach, we are free to choose the common value

$$w_{\alpha,M} = 1 \tag{7.2}$$

for all agents. This choice makes the individual wealth functions comparable and yields a monetary equivalent of wealth in money units. These can be summed up.

The usual economic measure of wealth is GDP, a measure of economic activity. (Indeed, it is well-known that GDP is *not* a good measure of wealth, because it includes many so-called 'defensive' expenditures, like security guards, prisons and garbage disposal, that clearly do not add to real wealth. At the same time, GDP omits other benefits, such as non-monetized child care, household services, social services and environmental services.)

However, we find it useful to consider the relationship between GDP and our proposed measure of wealth. GDP is defined as the sum total of all returns to labor (wages and salaries) plus returns to capital (interest, dividends and rents). From the producer side, this can be expressed as the sum of payments to labor by all economic agents, plus gross profits of firms (including costs of capital)

$$GDP = \sum_{\alpha} (\pi_{\alpha} + \phi_{\alpha}) \tag{7.3}$$

where π_{α} is the gross profit of agent α, and ϕ_{α} is the wage and salary bill for labor paid by the firm. Note that the sum of these two items for firm α constitutes the *value-added* (to purchased non-labor inputs) by the firm. In this case the index α is summed over all firms (*non-household* agents).

The sum of wages and salaries by firms can be equated to returns to labor (household income from work) while costs of capital to firms consists of dividends and interest payments to individuals plus direct investment by firms. Household income, in turn, consists of wages and salaries plus dividends, interest and rents.

Household income can simultaneously be equated to personal consumption expenditure C plus net personal savings S, which we assume is invested. Total investment I is therefore the sum of direct investment by firms plus net personal savings S by households. It follows that GDP is the sum of consumption C plus investment I. Total consumption is, of course, the sum of payments (price times quantity) for consumption goods and services, while investment is the sum of payments (price times quantity) for capital goods. It is now convenient to write

$$GDP = C + I = \sum_{k} p_k C_k + \sum_{i} p_i I_i \tag{7.4}$$

where I_i is the ith producer capital (investment) good, and C_k is the kth final consumption good. For later reference we note that capital goods may include inventories of raw and semi-finished materials, as well as machinery, equipment and structures. By the same token, consumption goods also include some durables (e.g. houses, clothes, furniture and automobiles).

Having said this, it is interesting to consider the *change* in total wealth from one period to the next. Comparison of wealth between time t and time t + 1:

$$Z(t+1) - Z(t) = \sum_\alpha (Z_\alpha(t+1) - Z_\alpha(t)) \tag{7.5}$$

Now the term on the right-hand side can be decomposed into monetary and material parts, by the help of *Equation 4.6* viz.

$$Z_\alpha(t+1) - Z_\alpha(t) = w_{\alpha,M}\left(\Delta M_\alpha + \sum_i v_{\alpha,i} \Delta X_{\alpha,i} \right) \tag{7.6}$$

where $w_{\alpha,M}$ is the subjective value of money to agent α, and $v_{\alpha,i}$ is the subjective value of the ith good to agent α. As noted above (*Equation 7.2*) we now assume that the value of money for all agents is the same.

The money change term ΔM on the right-hand side of *Equation 7.6* is profits (in the case of firms) or savings (in the case of households). Profits can be equated to revenues (from sales, subsidies from government in the case of government agencies, and/or from gifts or endowments in the case of universities or hospitals or other not-for profit organizations) minus the total payments for raw materials, intermediates, and purchased services, including wages and taxes. Savings (in the case of households) consist of income less expenditures for consumer goods. (Household savings can be negative, thanks to credit.) The material term ΔX consists of changes in physical inventories of raw and semi-finished materials, and durable capital goods (in the case of producers) and accumulations of consumer durables (in the case of households).

Stock changes can be divided into two categories, according to whether the changes result from market or non-market processes. We can classify as market processes those which are accounted for in the GDP calculation. Taxes and subsidies are included with more conventional market processes. All others are non-market processes, by definition. Non-market processes include physical deterioration, catastrophic losses (e.g. due to storms or floods), losses due to theft or vandalism, and so forth. For convenience we assign depreciation to the non-market category. Rewriting *Equation 7.6* we get

$$Z_\alpha(t+1) - Z_\alpha(t) = w_{\alpha,M}\left(\Delta M_\alpha + \sum_i v_{\alpha,i}(I_{\alpha,i})_m + (\Delta X_{\alpha,i})_n\right) \qquad (7.7)$$

where the subscript m indicates that the term reflects changes due to market activities such as investment, while the subscript n indicates changes in stocks that result from processes that occur outside the market. Non-market changes with a positive sign would include discoveries, and products of unpaid labor, including products of subsistence farms, hunting and gathering activities, and do-it-yourself (DIY) projects. Changes with a negative sign include accidental losses due to storms or fires, natural deterioration, corrosion, erosion and pollution damage to common property resources, including climate and bio-diversity.

The link between changes of Z and GDP is now straightforward if (and only if) distributional effects are neglected. Disregarding the differences in subjective valuation, as noted, the aggregated money stock changes of all the agents, individually, will then be equal to the change in aggregate money supply of the economy as a whole during the period. Money circulates. Profits and savings in aggregate cannot simply accumulate for long, despite the legendary French peasants who kept gold under their floorboards. To a very good approximation, all savings and all net profits pass through the financial system and eventually become investments.

We can now express GDP in terms of investment I, consumption C and the price and value variables introduced in earlier chapters, viz.

$$GDP = C + I = C + \sum_\alpha \sum_i v_{\alpha,i} I_{\alpha,i} + \sum_\alpha \sum_i (p_i - v_{\alpha,i}) I_{\alpha,i} \qquad (7.8)$$

The investment term I in *Equation 7.8* has been decomposed into two parts. The first part, which is the sum over all agents and goods of subjective values times quantities invested represents the value of the investment to the investor(s). The second term, which is the sum over differences between actual costs (prices paid) and subjective valuations of investment goods is something like 'investor's surplus' (analogous to consumer surplus). It can be interpreted as *the subjective net present value to investors of the expected future profits that will be earned as a result of the incremental investments.* The stock market would reflect this term as an increase in the market value of shares, less book value, during the period. For simplicity, we denote this term, hereafter, as Λ.

Similarly, it is convenient to introduce the symbol Γ as the sum over individual money stocks. Thus $\Delta\Gamma$ can be interpreted as the change in the aggregate money supply from one period to the next. Substituting from

Equation 7.7 and assuming all agents share the same subjective value of money w_M (as in *Equation 7.2*), we obtain:

$$\Delta Z = GDP - C + \Lambda + \Delta\Gamma + \Delta Z_n \qquad (7.9)$$

where ΔZ_n is for wealth changes that result from processes that occur outside the market as specified in *Equation 7.7*.

The fact that current consumption subtracts from the increase in wealth from period to period is not counter-intuitive, of course. The last term in *Equation 7.9* incorporates the combined effects of non-market processes, both those resulting from unpaid labor and those resulting from accidents and damage to common property resources, as discussed above, following *Equation 7.7*.

7.10   CONCLUDING COMMENTS

It would be wrong to end this chapter without pointing out, in the most emphatic manner, that GDP per capita is not really a measure of welfare. That it is commonly (mis)used in this sense by politicians and business and financial leaders does not excuse the practice. GDP is a measure of economic activity, nothing more. As has been pointed out by numerous critics, an automobile accident resulting in injuries requiring hospitalization, or a fire that destroys a house or a beautiful forest calls forth responses that add to the GDP but do not increase welfare.

There have been serious attempts by economists to resolve this difficulty. Perhaps the first was by Tobin and Nordhaus (Tobin and Nordhaus 1972). In order to address the question whether GDP/capita growth was 'real' (i.e. welfare-increasing), or not, they compared trends in GDP/capita as published *vis-à-vis* a modified version in which they subtracted items that were obviously defensive in nature. Since the two trends tracked each other reasonably well for the post-war period up to 1970, Tobin and Nordhaus concluded that GDP/capita was an acceptable proxy for welfare, at the national level.

This conclusion has been severely challenged in the 1990s, especially by Daly and Cobb (Daly and Cobb 1989; Cobb and Cobb 1994; Jackson and Marks 1994). In effect, these authors revised and extended the earlier work of Tobin and Nordhaus, into more recent decades and several additional countries, including Japan, Germany and the UK. The bottom line from these more recent studies is that the welfare-adjusted GDP no longer tracks standard GDP closely. Whereas GDP continues to increase in most parts of the world, it appears that per capita welfare has not been increasing much, if at all, since the 1970s.

Sociological studies seem to confirm this finding. It seems that most people judge their welfare mainly in relative, rather than absolute terms (i.e. relative to others in their acquaintance). If everyone's wealth increases by the same amount, relative welfare remains equal. Clearly, any trend towards egalitarianism (e.g. the Scandinavian model) is likely to increase the welfare of the lower income people, who are the majority, relative to those at the top, whereas a trend in the opposite direction (such as the US has experienced since the 1980s) tends to reduce relative welfare for the majority. It follows, incidentally, that conspicuous consumption by the rich has the perverse effect of making everyone else feel worse off, thus lowering real utility at each income level.

Sadly, it is not possible to deal with these issues fully in a book such as this. We can only acknowledge that consumption issues in the aggregate economy cannot be divorced from income distribution, factors affecting – and changing – preferences, and the increasing disconnection between consumer welfare and income. These issues deserve a research program of their own.

NOTES

1. Here we must recognize a fundamental departure: microeconomic theories (including ours) assume path-independence, in order to establish the existence of a utility function (in the standard theory) or a Z-function, in our case. This assumption is clearly at odds with reality, and can be allowed as an approximation only in limited circumstances. On the larger macro-scale, the assumption cannot be so easily accepted.
2. There have been a number of famous historical examples of bubbles, starting with the seventeenth century tulip mania in Holland (1637), the Mississippi Bubble in France (1718–20) and the nearly simultaneous South Sea Bubble in England (1720), and numerous examples since then, including the US stock market bubble of the late 1920s which ended with the Great Crash of October 1929, the Tokyo real estate bubble of the late 1980s and most recently the 'dot.com' bubble of the late 1990s, which ended in March 2001.
3. 'Spillovers' are beneficial externalities, as when the transactions between two economic agents are beneficial to a 'third party', even though the latter is not involved in the transaction. Probably the most important examples of spillovers occur in the knowledge domain, as when a discovery or invention sponsored by party A (who pays for it) happens also to be useful to, or solves a problem for, party B. For instance, the development of radar in World War II made microwave communications and microwave ovens possible.
4. This situation is changing rapidly, as more and more women (in cities) take jobs outside the home and hire other women to care for children or perform other domestic tasks.
5. For some context see also (Hobsbawm 1975).
6. As it happens, the term 'energy' as normally used in this context is theoretically incorrect. The quantity of interest is really '*available energy*' or '*exergy*'. Our choice of unit hereafter will be *exergy* content. This term is unfortunately not familiar to most people, and a full explanation would take us rather far afield. For the present, it will be sufficient to think of exergy as 'useful energy'. (The essential point being that not all energy is useful.) As it happens, the term exergy can also be defined for non-fuel resources and used as a measure of quantity for such resources. Thus it is uniquely valuable for comparative analysis of material flows. The detailed arguments are given in other publications, however.

7. Robinson Crusoe was the name of a fictional character who was shipwrecked on a deserted island and forced by circumstances to feed, clothe and house himself utilizing only the resources provided by the island.

8. The drivers of long-term growth: knowledge, technological change and radical innovation

8.1 INTRODUCTION

In the previous chapters of this book we have articulated in some detail how individuals (and firms) belonging to the subspecies *H. Economicus* constantly try to increase their wealth, subject to the AAL rule. The last chapter discussed the problem of aggregation, and the transition from a microeconomic focus to a macroeconomic focus on the system as a whole. However, while technical progress is, in some sense, a reflection of additions to human knowledge, and while learning and adaptation, in particular, take place at the level of individual economic agents, it is not possible to treat technical progress at the system level as an aggregation of small increments to knowledge.

The core problem is that, in fact, learning and incremental change do not explain radical innovations that change the structure of the economy. Improvements in gas light did not, and could not, explain the advent of electric light. Nor did Edison's incandescent electric light with its DC generator necessarily bring forth Tesla's inductive motor and the three-phase power distribution system, the Hall-Heroult electrolytic process for making aluminum, or Moissan's electric furnace; nor did these breakthroughs necessitate Marconi's radio-telegraph, the vacuum tube diode, the superheterodyne circuit, TV or the ENIAC electronic computer. These subsequent developments can be regarded, however, as 'spillovers' from Edison's innovation. Similarly, the vacuum tube did not evolve into a transistor, nor did the abacus or the mechanical calculator morph into an electronic computer, and the applications that followed. But the latter are spillovers that have accounted for much of the economic growth of recent years.

The crucial importance of such spillovers for growth arises from the fact that incremental improvements of existing devices and methods all tend to approach limits. As limits approach, incremental improvements gradually become more costly to achieve; in other words, the returns to research and development decline. Meanwhile, notwithstanding the standard assumption of *non-satiation*, real markets for all types of existing goods do eventually approach saturation. The 'garage sale' phenomenon of recent decades illustrates how much superfluous, but not yet worn-out, clothing, furnishings, appliances, toys and sports equipment now exist in attics, basements and garages around America and Europe. Needless to say, this is bad news for the manufacturers and distributors of those goods. Satiation was an almost unimaginable prospect to the nineteenth century economists, but in some markets, at least, it can no longer be dismissed as a prospect as distant as the cooling of the sun, or even the end of the age of oil.

In a world of limits, both to technical improvements and to market demand, there is no way the gradual incremental change mechanisms at the micro-level, described in earlier chapters, can explain the radical innovations that have in the past, and may in the future, account for the creation of new products, structural change and long-term economic growth.

To be sure, Joseph Schumpeter gave a partial explanation of radical innovation long ago in his Ph.D. thesis (1912) and again later in his important book, *Theory of Economic Development* (Schumpeter 1912, 1934). To summarize a long story in a few words, Schumpeter's key idea was that firms are motivated to undertake radical innovations in order to make 'extraordinary' (i.e. monopoly) profits resulting from exclusive ownership or access to the new product, process or 'combination' (his word).

The extraordinary profitability of monopolies was well understood, in Schumpeter's time, and had been amply illustrated by such examples as Western Union, Pennsylvania Railroad, Standard Oil and Eastman Kodak. Indeed, Wall Street financiers after 1890 and through the 1920s were busily assembling other would-be monopolies in order to take advantage of this superior profitability. Examples of firms created by mergers in that period included AT&T, General Electric, US Steel, General Motors, IBM and RCA. Schumpeter himself was convinced that large firms were best suited to undertake major innovations, precisely for this reason. He was wrong about that, as it happens, because it has turned out that large firms, in general, tend to be more risk averse than small ones.[1] And this is a crucial point.

An important feature of technological knowledge that is pertinent to Schumpeter's thesis is that, with very rare exceptions, it can only be monopolized by an innovator for a short time, perhaps a few years. Eventually the knowledge diffuses and becomes available to competitors. (In fact, the inevitability of knowledge and information diffusion can be regarded as still another irreversibility.) Nevertheless, the 'extraordinary' profits obtainable from the initial monopoly may be one of the drivers of Schumpeterian innovators. That is to say, the hope of extraordinary profits was an incentive for risk-taking.

Did the risk-taking pay off? In the later nineteenth and early twentieth centuries it may have done. We have little evidence, one way or the other.

However recent evidence suggests that Schumpeterian profits over the past 30 years or so have been very small, at best.[2] Other incentives to innovate have almost certainly been at work. But it is certain that the benefits to the rest of the society from the eventual spillovers far outweigh the benefits captured by innovators. These spillovers almost certainly account for longer term economic growth and the continuing growth of total factor productivity (TFP). This idea is central to the one strand of the so-called 'new endogenous growth theory' (e.g. Romer 1986, 1990; Lucas 1988).

Having said this, it remains unclear how the AAL rule can be reconciled with major or radical innovations that are almost invariably extremely risky undertakings in their early stages. To put it bluntly, when it comes to radical change, the odds against success are extremely high, even for a well-funded and managerially competent start-up with a really new idea. This is one of the reasons why big companies with competent financial and accounting staff capabilities almost never invest in high risk development enterprises. The odds facing an individual without access to resources are nearly astronomical. In terms of *rational expectations*, the prospects for any would-be innovator are clearly negative. So, why do eager inventors keep trying?

Part of the explanation, of course, is that ambitious young inventors do not realize the difficulties they face. They are, in the words of the song, 'cock-eyed optimists'. But part of the explanation has to be that inventors and innovators are not rational in the economic sense. Rather than avoiding losses, it would seem that they seek risks and ignore losses. For society as a whole, of course, the situation is reversed: society as a whole benefits enormously from the losses sustained by its most imaginative and creative members.

8.2   THE BEHAVIORAL BACKGROUND

A behavioral characteristic with an evolutionary origin is the desire to learn about the world one lives in. Human curiosity (sometimes called 'monkey curiosity') probably does not need explanation in economic terms, since human curiosity preceded economic relationships. It is a behavioral characteristic common to most higher species of animals, with obvious evolutionary survival benefits. One sees it exemplified most clearly in the exploratory behavior of puppies or kittens, between meals and naps. Young children explore in a similar manner. The survival benefits are obvious: the more an individual organism knows about its environment, the more easily it can avoid dangers and find shelter or food.

For firms much the same incentives to explore are applicable, although the environment is different and mostly non-physical. Of course, exploration in the physical domain is an important aspect of the extraction industries, especially mining, oil and gas. But exploration in a different domain is no less important for other firms. Market research is a systematic exploration of the parameters of demand for products and services. R&D can then be regarded as exploration of the possibilities for supply of products and services, whether by changing the characteristics of the product (or service) or by improving the production method.

Risk-taking is a different sort of behavioral characteristic. Willingness to take risks in the face of danger – known as the 'fight or flight' instinct – is apparently an important survival quality for all animals, including humans. The thesis has been proposed by some anthropologists that risk-taking in the face of obvious and immediate danger to life is instinctive, and that it explains the human propensity for gambling. The question remains whether risk-taking in life-threatening situations explains willingness to take the sort of risk that many inventors have chosen, namely the overwhelming probability of spending many years of deprivation, frustration and obscurity against the (statistically) slight chance of ultimate vindication and wealth. The skeptical counter-argument, of course, is that most inventors are deluded fools who do not know, or simply refuse to acknowledge, the real odds against them. A further counter-argument is that an inventor or entrepreneur who has already invested time and money on a project becomes more and more unwilling to cut his/her losses as time goes on. (This is closely analogous to the well-documented reluctance of investors to sell shares at a loss, in hopes of a later recovery, even though professional investment advisors point out that what is lost is lost, and that the money that remains might be better invested elsewhere.)

Propensity for risk-taking combined with curiosity (and luck) probably played a major role in driving early innovations. The earliest examples were probably weapons, initially bones used as clubs, and later sharp stones attached to sticks with thongs from animal skins. The deliberate production of arrowheads and spearheads from flint – a stone particularly well suited for the purpose – must have followed almost automatically. But the invention of the bow and arrow was a radical innovation by any test, and a stroke of genius. The refinement of this technology, and the development of skills to use it effectively, must have taken tens, if not hundreds, of generations. This was an example of learning-by-doing.

The first use of fire to cook meat was almost certainly another case of serendipity, resulting from the discovery of partially cooked but edible remains of animals trapped by wildfires. The capture of fire itself in the form of burning brands and its use for warming cave shelters and driving off predators would have followed naturally, especially in the glacial periods. The use of inedible animal skins for other purposes, especially protection against cold, must have also followed naturally soon after the early hunters began bringing their trophies back to the cave and the campfire for cooking and distribution. But the art of cooking developed over millennia since then was, again, a case of learning-by-doing. Finally, the first use of fire to harden bits of wet clay to create crude pots was also probably an accidental discovery, while the refinement of the technology over hundreds of generations is yet another example of learning-by-doing.

In short, while the subsequent improvements owe much to learning and incremental improvements, the original breakthroughs were likely attributable to something else, either luck or genius, combined with some propensity to take risks. The first man to defend himself against a wild boar or a leopard with only a fire sharpened stick took a big chance. He probably did so because it was an emergency of some sort. Perhaps he couldn't keep up with the other hunters, due to an injury. Nevertheless, it was a risky choice. The first man to use a bow and arrow successfully for defense, or for hunting, probably had to break the traditional and established 'rules' governing hunting to do so. He risked his status in the tribe, if not his life.

We noted earlier that a member of *H. Economicus* never buys a lottery ticket or bets on a horse race. The reason is obvious: the expectation value of such purchases is negative. For exactly the same reason, a moderately risk-averse investor will almost never invest in a new restaurant or small start-up company if he or she knows the real odds against success. Those who do invest in such enterprises are usually close relatives or they are misinformed. Despite the tales of spectacular success in a few well publicized cases, even professional venture capital firms also rarely invest in true start-ups, because so few succeed. They greatly prefer second or third stage financing for those start-ups that survive the first year or two, and even so, they expect to lose most of their investments. It is the occasional big winner that justifies the rest (and attracts the punters).

Who are the investors in truly radical new technologies? They are, above all, the inventors themselves and their very close relatives. Most of these individuals are passionate believers in their own genius, unaware of the obstacles and focused exclusively on the dream of ultimate success. Great wealth is often part of that dream, to be sure, but the dreamers are hardly rational utility maximizers, or even followers of the AAL rule. More often than not it is fame or vindication they seek, above all else. The point we are making is that while such risk-taking inventors clearly belong to *H. Sapiens*, they are not members of *H. Economicus* as we understand that group. But it must be acknowledged without question that these unusual and 'irrational' people have accounted for a remarkable share of the important new products and new processes that have driven economic growth over the past two centuries and more.

Violation of the commandments of the AAL rule may be a necessary condition for success as an innovator, but it is scarcely sufficient. Many will take chances. Few will succeed. The remainder of this chapter deals with other prerequisites of radical innovations. However, it is as well to acknowledge here and now that radical innovation – hence long-term economic growth – cannot be endogenized within our AAL-based theory, still less the standard neoclassical utility maximization paradigm.

8.3   STANDARD MICROECONOMIC THEORY OF TECHNOLOGICAL CHANGE

Stepping back from the behavioral issue, there are two ways in which innovations can contribute to economic growth. The more common, but less exciting, is when a new product replaces an old one in a given market niche, at a lower price. In this case consumers can buy more of that product, or they will have a little extra money to purchase other goods and services, or both. Increased sales drive increased investment in capacity, or in R&D, yielding improved performance, economies of scale, increasing production experience, further cost reductions resulting in further price reductions and so on. This is a primary 'engine of growth'. It is not sufficient to explain actual economic growth, however, because the positive feedback cycle encounters declining returns. Declining returns can only be compensated for by the introduction of radically new products and technologies, as noted above.

The ordinary economic incentives for doing most kinds of research are straightforward: any knowledge increment that increases operational effectiveness or adds value to a product or service adds to the individual or firm's competitive advantage in the market and thus to its competitiveness. A worker can add to his or her value to an employer by learning a new skill. The learning process may be almost cost-free (i.e. learning-by-doing or learning-by-using). On the other hand, it may require some investment in leisure time, or a monetary investment in formal training. New or honed skills add to future earning power. As noted in a previous chapter, added knowledge or skills can add to an individual's 'value', as expressed in terms of current borrowing power without adding to personal welfare or satisfaction. While credit is not actually a form of money (because it must be repaid eventually with money), it increases liquidity, which is commercially valuable in itself. Most important, it enables a borrower to spend – or invest – before earning the money to repay.

The same argument applies to producers (firms), though in a slightly different way. The monetary value of a firm is usually more than the value of its material assets and money in the bank. In some cases, such as biotechnology start-ups, most of the value is embodied in the skills and knowledge of the employees, plus their formal intellectual property (patents, patent applications, proprietary software, etc.). In other cases, much of the value is 'goodwill' including trademarks and reputation. Just as the essence of money is trust in institutions, the same can be said of trademarks. Reputation for honest dealing and prompt payment establishes commercial credit, which is very important in periods of uncertainty.

Evidently skills, knowledge and ideas potentially have monetary value for firms, mostly in terms of the ability to create new products (or at least useful variants of old ones) and bring them to market. A considerable body of research by Edwin Mansfield and others has shown empirically that R&D investment can be much more profitable than ordinary production and sales activities (Mansfield 1965; Mansfield et al. 1977; Mansfield 1981; Mansfield et al. 1983). The reason, presumably, is that R&D can yield product improvements, or new products, that differentiate an existing product or product line from its competitors and thus provide a quasi-Schumpeterian advantage in the marketplace. This advantage can be measured, at the firm level, in terms of the net difference between the market valuation of tradeable shares and so-called 'book value' (tangible assets plus money in the bank and receivables, minus debts). This differential clearly reflects future earnings prospects and, therefore, can be regarded as a monetary measure of a firm's proprietary technology.

From the 'standard' macro-perspective, the core theory of technological change is usually termed 'induced innovation'. This theory has been elaborated qualitatively in several books by Rosenberg (for example, Rosenberg 1969, 1976, 1982) and, in more theoretical and mathematical terms, by Binswanger and Ruttan (Binswanger and Ruttan 1978). Here the fundamental idea is that scarcity 'induces' innovation, albeit the incentives at the micro-scale are not explained. For example, economic historians have argued persuasively that, in the nineteenth century, the US was short of labor (compared to Europe), but had plenty of good land and fodder for horses. This combination made horse-drawn harvesters and other kinds of agricultural mechanization more profitable to utilize in the US than in Europe. This seems to explain why many innovations in that field, such as the combine harvester (and later the tractor), were invented and utilized first in the land-rich but labor-scarce US.

The theory of induced innovation applies specifically and especially to the impact of natural resource scarcity – real or perceived – on economic growth. Modern resource economics began with a famous paper on the economics of exhaustible resources by Harold Hotelling (Hotelling 1931). However, the possible contribution of natural resource inputs to economic growth (or to technical progress), was not considered seriously by economists until the 1960s, especially due to the path-breaking study by Barnett and Morse (Barnett and Morse 1962) sponsored by Resources for the Future (RFF). The message of that study, which relied heavily on long-term price trends for exhaustible resources, was that scarcity was not an immediate problem, nor likely to be one in the near future, thanks to technological progress in the realm of exploration and extraction, and the substitution of alternative materials whenever a scarcity threatened.

This optimistic conclusion was briefly challenged by events of the early 1970s, including the 'energy crisis', the rise of OPEC and partly in response to the Club of Rome's *Limits to Growth* report (Meadows et al. 1972). Economists responded immediately with a number of papers disputing the *Limits to Growth* methodology and conclusions, (e.g. Solow 1974; Stiglitz 1974; Dasgupta and Heal 1974). The potential for substitution was particularly emphasized by Goeller and Weinberg (Goeller and Weinberg 1976).

In more recent applications of the standard theory, possibly influenced by the 'limits' debate, resource consumption has been treated as a *consequence* of growth and not as a factor of production (Solow 1986, 1992; Smith and Krutilla 1979). This assumption is built into virtually all textbooks and most of the large-scale models used for policy guidance by governments. We argue *a priori* that the assumption of uni-directional causality is false: that energy (exergy) consumption is as much a driver of growth as a consequence. However the full development of this thesis cannot be undertaken here.

The microeconomic theory of change (hence growth) presented earlier in this book, especially Chapter 6, postulates learning from experience. We have previously discussed learning mainly in the context of risk management, time preference (discounting) and bargaining negotiation. However, learning also involves other kinds of knowledge accumulation. Learning-by-doing (which can be 'embodied' in the design of capital equipment as well as workers and organizations) is one of the classic mechanisms to increase labor productivity and cut production costs. An early, and still influential article on this process appeared in the economics literature four decades ago (Arrow 1962). Indeed, economists, by and large, have not gone far beyond Arrow's analysis. Many economists are still characterizing technological change at the firm level as a kind of random process, analogous to mutation in biological evolution or – perhaps a better analogy – to purchasing lottery tickets (in the form of R&D expenditures), except that R&D pays much better, provided it is narrowly focused on small incremental improvements rather than radical innovations.

For a very simplified theory of growth and change in a single sector this may suffice. But a number of important questions arise that the simple theory and models cannot answer. For example: if the forces driving discovery, invention and innovation are economically determined (as we presume), then why should technological progress be so discontinuous?[3] How can economic theory explain the enormous (and sudden) shifts in R&D from one field to another? Why are some fields of research abandoned for decades, then suddenly 'rediscovered?' Most important of all, is it possible to make plausible projections of the rate and direction of technological change into the future?[4] This question concerns us in the next section.

8.4   THE DELIBERATE SEARCH: BARRIERS AND BREAKTHROUGHS

The earliest radical innovations were probably partly due to luck and curiosity, but curiosity cannot account for the deliberate and systematic search for new combinations and configurations, to overcome a perceived barrier and solve a specific problem. Archimedes *c*.200 BC) was working for the king of Syracuse, who was seeking a way of proving whether the gold coins from his mint had been adulterated with silver, or not. The answer (Archimedes' principle) was inspired by his bath. Newton is said to have conceived the law of gravitation as a result of an apple falling on his head. But, in both cases, the event that triggered the important insight was actually the culmination of an intellectual search process of some sort.

It is possible (and interesting) to document the history of technology, especially radical innovations, in terms of search processes triggered by war, scarcity or some other crisis. It is banal, but no less true, to say that 'necessity is the mother of invention'. Much less banal is the observation that only a few of the major inventions and innovations of the past have contributed, via the 'spillover' effect, to the creation of significant new products and industries.

The steam engine was one of those, of course. It led directly to the development of railways and steamships, and indirectly to the development of gas light. The Bessemer process in steel-making was another example. It lowered the cost of steel so much that steel finally became a major construction material. Of course it contributed dramatically to the spread of railways in the US and elsewhere, but it also enabled steel bridges, steel-hulled ships, sky-scrapers, steel pipe, steel barbed wire fencing, modern guns and a host of other products. (The incidental fact that the Bessemer process was quickly replaced by the open hearth process is irrelevant.)

Electric power may have been the champion example of an innovation resulting in spillovers enabling (if not creating) new products and industries. It was originally developed for lighting purposes, but applied almost immediately to electric motors, which were applied to trams, railways, elevators, factories as well as pumps and compressors that enabled vacuum technology, refrigeration and air-conditioning. Electric power also allowed electrolysis, which resulted in commercially viable aluminum, chlorine, magnesium and metallic sodium, potassium and phosphorus. Electric furnaces permitted the production of acetylene (from calcium carbide), synthetic abrasives like silicon carbide (carborundum), and refractory metals including nickel, chromium, cobalt, and the alloy stainless steel. Synthetic abrasives made possible high-speed grinders and drills, without which some key components of the internal combustion engine (for example, the crankshaft) could never have been mass-produced. Finally, electric power is an essential prerequisite for telegraphs, telephones, radio, TV, radar and the whole range of applications of electron-ics, including computers.

The breeding process has continued through the twentieth century. The transistor – in turn – has enabled portable radios and replaced vacuum tubes in virtually all applications. As computers became more complex, the transis-tor morphed into the integrated circuit or IC (the 'chip'), which evolved from large-scale integration (LSI) to very large-scale integration (VLSI) and the microprocessor. These innovations, radical in themselves, made computers much more powerful and more reliable, while also spreading into many other domains, from cellular phones to machine controls. Since 1980 software development and the shift from analog signal processing to digital signal processing have overtaken the hardware development as the leading edge of progress. The convergence of telecommunications and computer technology, made possible by digitalization, created the Internet and may be generating yet another wave of spillovers capable of creating new industries or revolu-tionizing old ones. Indeed, those unrealized (as yet) spillovers may provide the impetus to drive future economic growth, now that many of the older industries are confronting problems of market saturation and overcapacity.

8.5   TOWARD A NEW QUANTITATIVE GROWTH THEORY

Perhaps the major point made in this chapter is one that mainstream econo-mists would not have troubled to make at all, namely that standard economic theory does not explain radical technological innovations, nor does it explain long-term economic growth and structural change. Standard economic theory has essentially finessed this gap in the theory by postulating that continuous multi-factor productivity (MFP) improvement is simply a fact of nature. Keynes allegedly called it 'animal spirits' and left it at that. We agree that it cannot be explained entirely within the neoclassical theory, but we do not agree that MFP affects all sectors equally, or that it can be expected to continue indefinitely. The point of difference with more conventional theories is that we do not view technological change (TFP) as automatic and cost-free. We acknowledge the importance of spillovers, but we believe that most 'breakthroughs' do not have significant implications outside the fields where they were made. This applies, for instance, to medicine and health, where the admittedly great progress that is being made in the development of new drugs has little if any impact on other sectors of the economy.

The reason we have emphasized the importance of radical Schumpeterian innovations, and their critical role in driving long-term growth is to underline several points: (1) that such innovations are *not* random events, (2) that technological change does *not* occur uniformly across all fields at once, but tends to be rather narrowly focused on particular areas at any given time and (3) that each major breakthrough depended upon earlier ones, whence there appears to be a natural ordering or sequence of breakthroughs. This natural order is determined partly by geopolitical events, partly by the distribution of natural resources among nations, and recently in large part by the laws of nature and the properties of real materials.[5]

The latter point deserves much more extended treatment, which we cannot undertake within the scope of this book. However, it supports – from a slightly different direction – our stated conclusion that radical technological change is inconsistent with rational behavior, whether defined in terms of utility maximization or the AAL rule, hence cannot be predicted within the framework of standard economic theory.

It is our contention, to be elaborated in a subsequent book, that the basic driving force for technological change and economic growth since the mid eighteenth century has been the increasing substitution of machines for animal and human labor. This substitution has been characterized by increasing efficiency (hence decreasing cost) of mechanical work *vis-à-vis* animal or human work. The many radical innovations involved in this trend toward increasing efficiency and lower cost, including electrification, have also occurred in a natural order (with a few possible exceptions). It will be argued that this sequence of innovations, and the resulting increase in resource conversion efficiency, accounts rather well for the economic growth of the US (and other countries) since the beginning of the twentieth century.

NOTES

1. This is not to say that large firms were never innovative. In fact, most of those listed (except US Steel) were quite innovative at that time. This was partly due to competition. None of the mergers except the US Steel merger, was comprehensive enough to eliminate effective competition. For example, GE was in a continuing battle with Westinghouse, a smaller and more agile enterprise with rights to the inventions of Nicola Tesla. GM was in a continuing battle with Ford and Chrysler (among others), AT&T was challenged in the long-distance market by Western Union and Marconi. IBM was challenged in the punched card equipment field by Remington-Rand, and so on. Nevertheless, the predominance of small firms, individual inventors and spinoffs from university laboratories in the history of technology has been well documented (especially by Jewkes et al 1958).
2. A recent study estimates that innovators have only been able to capture 2.2% of the social surplus created by their innovations during the study period 1948–2001(Nordhaus 2004).

3. In this connection, there is a fascinating literature on the so-called Kondratieff (long-wave) cycle (e.g. Kondratieff 1926; Rostow 1975; van der Zwan 1979; Clark et al. 1983; Freeman 1983; Mansfield 1983; Perez-Perez 1983; Rosenberg and Frischtak 1984; Ayres 1989).

4. There was a major attempt to address this exact question in 1962 (Nelson 1962). However, while the conference (and the resulting book) included a number of ground-breaking papers, the authors of the papers included had little to say on either the rate or direction of technological change *per se*.

5. The historical sequence of breakthrough technologies has been partially determined by the properties of materials. For example, the order in which metals found important uses is largely determined by their availability in recoverable form and their melting points. Gold, silver and copper were found in pure nugget form that could be shaped easily by heating and hammering thousands of years ago. Copper and tin ores were found near each other (in Cornwall, for instance) and this fact encouraged the development of bronze. Pig iron required somewhat higher temperatures than copper and zinc, but temperatures that were still achievable in the middle ages. But the forming of pig iron into steel (to make good quality swords, for instance) was a very difficult process, involving a great deal of labor. Pure iron or steel could not be produced in molten form until Cort's puddling process was invented in the eighteenth century and it could not be produced efficiently in large quantities until the Kelly-Bessemer breakthrough in the mid nineteenth century. Other so-called refractory metals could not be melted, refined or alloyed until the advent of Moissan's electric furnace, toward the end of that century.

Appendix A: Money and credit

A.1 MONEY, CREDIT AND BANKS

Barter and bargaining between individuals has probably existed as long as humans walked upright. All organized societies of which we have any knowledge engaged in trade. In most cases trade was mostly barter, but in time barter was supplemented by at least one 'established' medium of exchange. The Aztecs used cacao beans. Berbers used salt. Coconuts and ivory were used in some places. Many societies bordering the Indian Ocean used cowrie shells. Mountain people sometimes used olive oil or sheepskins or goatskins. And, of course, many societies used metals, ranging from gold and silver to bronze and even iron (Weatherford 1997). The first recognizable gold and silver coins, actually produced by a mint in standard sizes (and stamped with a lion's head), were introduced around 560 BC by the King of Lydia, in Asia Minor. This innovation so facilitated retail trade and local production (mainly of cosmetics) that Lydia quickly became a regional trade center. In fact Croesus' wealth is legendary.[1] Croesus' monetary innovation quickly spread throughout the Greek world, and later, the Roman world.

The second major innovation in the history of money came 1500 years later. The innovation was credit. The first 'bank' offering credit (but only to the feudal nobility) was the Order of the Knights of the Temple of Solomon, later known as the Templars. The order was founded in Jerusalem in 1118 AD to defend the Christian enclaves in the Holy Land, and the sea and land routes thereto. The Templars failed to save Jerusalem, but they performed financial services for crusaders, including loans, mortgages and even managing the royal revenues of Philip II of France after 1190 (Weatherford 1997). In 1314, King Philip IV or France ('the fair') arrested the Templar leaders on trumped up charges, and (with the complicity of the Pope) crushed the order hoping to confiscate all of its wealth.

The end of the Templars did not end the need for banking services. The need was met in the fourteenth and later centuries mainly by Italian merchant families in Florence, Venice, Genoa and Milan. These merchants offered their services to anybody (who could pay) and they found a legal way around the Church's injunction against usury. Their innovation – which converted merchant traders into banks – was the 'bill of exchange'. This device allowed a client to deposit money (in coinage) at a branch in one country and obtain equivalent funds, in coins of another realm at another branch of the bank at a later time. Needless to say, the bank charged for this exchange service. But since different currencies were involved, the fee was not 'interest' according to canon law.[2] Soon these bills were themselves being exchanged as portable credit, and became the first paper money.

The merchant-bankers relied on their reputations for honesty and fair dealing to attract clients. The bills of exchange were ultimately convertible into coinage, and coinage was trusted insofar as it was based on gold. For centuries, gold was widely regarded as a 'store of value' (and still is so regarded by some). However, since the end of the seventeenth century there has not been enough gold in the world – except during brief periods after major discoveries (such as California in the 1850s and South Africa in the 1880s) – to satisfy the burgeoning demand for money and credit.

In the early eighteenth century a Scottish economist and banking innovator named John Law was the first to propose national banks and paper money backed by other assets as a supplement to coinage. His specific scheme was flawed and failed spectacularly, but the basic idea was sound.[3] During the next two centuries the problem of gold scarcity was not solved by any general reform, but the shape of the solution evolved gradually and piecemeal from a series of experiments, resulting from local crises. National banks were created in the most advanced countries, and they, in turn, 'created' paper money by fixing an official price for gold, or for another currency backed by gold, and using a national gold reserve as support for it. Merchants trusted the paper money because they believed in this promise and they believed that gold was an ultimate source of value.

However during the two centuries since John Law's innovations it became increasingly clear that the backing for paper money was not really gold so much as the idea, or symbol, of gold. During the nineteenth century the British pound sterling was the real international standard of value, and the basis for international trade. After World War I, the US dollar took over this role. Yet, until 1933 the paper money issued by the US Federal Reserve Bank could still legally be exchanged for gold. Gold was still tied officially to the dollar (by a fixed price) until 1971, when the gold standard for trade purposes was finally abandoned by President Nixon. Since then the value of paper money has been officially backed (as it always had been unofficially backed) only by the citizen's trust in the economic system and the financial probity of the government itself.

The essence of money and credit is trust. It was no accident that many banks have used the word 'trust' in their names. The depositor must trust the bank to repay. In fact, it can be argued that commercial morality (recall the discussion in Chapter 1) co-evolved with the marketplace and the use of money. However, this is not our major point. The point is that trust, based on reputation, is an intangible.

A.2 THE MONEY SUPPLY

Money is a catalyst for trade. Lack of money (or lack of credit) is tantamount to lack of liquidity. Insufficient liquidity inhibits economic activity. If the bankers will not lend, consumers, merchants and manufacturers cannot borrow, and non-cash businesses must cease activity. Problems due to insufficient money (i.e. liquidity) have sometimes been acutely painful in the past. It was lack of liquidity that generated initial enthusiasm for the share offerings that underlay the Mississippi and South Sea 'bubbles' of the early eighteenth century. A scarcity of money in late nineteenth century rural America – actually a scarcity of credit – was widely assumed to be due to scarcity of gold.[4]

Checks backed by bank deposits do not increase the amount of money in circulation, but they do cut down on the amount of paper currency needed. Traveler's checks and bearer bonds (cashable by the bearer) are another form of circulating cash equivalent, which do not increase the effective money supply (since they must be paid for in cash) but are more convenient than gold or paper money. Commercial paper (so-called) backed by receivables is still another cash equivalent. Long-term loans (bonds or mortgages) increase the amount of money in circulation, although not the total quantity of money.

However other recent financial innovations do multiply the effective money supply by monetizing previously un-monetized assets. Credit cards are an important example. Shares in speculative ventures with future profit potential can have immediate market value. Shares representing claims on expected future profits are also a form of money, as John Law certainly recognized. (We discussed this issue tangentially in *Chapter 7* in connection with the interpretation of *Equations 7.6* and *7.8*, linking GDP and ΔZ.)

The fact that shares that are frequently traded on a major stock market are now effectively equivalent to cash, because they can be sold over the Internet in a matter of seconds, raises several critical points about market valuation. In the first place, stock prices can, and sometimes do, fluctuate widely from one day or one week to the next, even though the state of the firm has changed very little during that period, if at all. In theory, the market value of a share is the net present value of expected future earnings.

The notion of 'expected value' is slippery for several reasons. Obviously, very few investors have the information needed to make an intelligent assessment of future earnings prospects. In practice, simple 'rules of thumb' are normally utilized by investors. The so-called price-earnings ratio is the most familiar of these. It is curious – but very important – to note that the average (and norm) for this ratio has increased dramatically during the last (twentieth) century. In the late nineteenth century investors expected very large earnings, most of which were returned to investors immediately as dividends. Since large dividends were expected, firms in expanding industries were often unable to finance their own growth and were constantly running into financial crises.[5]

Today, most firms pay dividends of less than 5% (of the price of a share) and it is a rare industrial company that pays that much. Many pay 2%, or 1% or nothing at all. Investors, as a group, have increasingly sought capital growth potential rather than immediate earnings. This has induced firms to re-invest their earnings rather than distributing them to shareholders. On top of this, the stock market valuation norm in terms of price-earnings ratio has grown to very high levels in the so-called 'growth' sectors. The 1990s saw the advent of stocks, especially in biotechnology and so-called 'dot.coms', with high prices per unit of revenue and no earnings at all expected for several years into the future.

A related point is that many shares are now priced at levels far above their 'book' value, which is the realizable value of their cash and cash equivalents, real property and other saleable goods and equipment. What this means is that the market is putting a very large value on expectations. Yet banks and other lenders have been surprisingly willing to lend money based on such expectations. When a company of this sort is sold, this excess value is accounted for by the acquirer as 'goodwill'. If the market prices keep rising, wealth is 'created'. But if the market drops (as recently happened) the goodwill on the balance sheets must be written off, and the unwary acquirer can find itself with a large debt and insufficient real assets to secure it. Wealth is 'destroyed'.

The ease with which paper wealth can be created and destroyed by market fluctuations constitutes a major dilemma for governments today. On the one hand, it is not unreasonable for markets to attach value to intangibles, such as intellectual property, trademarks, and even research in progress. On the other hand, it is not plausible to assert that the excess value (above book) markets attach to many shares during periods of rising prices is attributable to a rational assessment of the present value of future earnings. On the contrary, what drives prices up during bull markets seems to be the expectation of higher prices *per se*. During bear markets, of course, the situation is reversed.

Reverting to the notion of money as a lubricant for trade, it is also true that, beyond a certain point, if the quantity of money in circulation increases too rapidly, this also inhibits trade. It does so essentially by making calculation difficult. Lenders cannot calculate appropriate interest rates for periods beyond hours or days, insurers cannot calculate premiums, credit dries up and only cash or equivalent goods are acceptable. In extreme cases (hyperinflation) too much money in circulation can cause the system to collapse. This nearly destroyed the United States in the first years after the Revolutionary War, and there have been several devastating hyper-inflations in the twentieth century, especially in Germany after World War I, in Hungary after World War II and in a number of South American countries.

A brief digression on *deflation* is appropriate here, given the indications that Japan has been experiencing it for some time, and Germany (as of mid-2003) may soon follow suit. The problem in brief is that, just as rising prices tend to increase demand, especially in property markets where most of the purchase price is borrowed, declining prices tend to reduce demand for the same reason. Consumers are especially disinclined to buy homes using borrowed money (mortgages) if there is a possibility that the property will be worth less on resale. To make matters worse, the macroeconomic stimulus potential available to central banks, namely to reduce interest rates, disappears when the prime rate falls below 1% or so. Interest rates cannot, by definition, go below zero. Once this happens, as it has in Japan, the only possible sources of stimulus to demand are fiscal policy (government spending) or tax cuts.

Given that too little money and too much money are both harmful, it follows logically that there must be an optimum money supply (liquidity) which maximizes the 'real' productivity of a society. The relationship between money supply and output (GDP) has evolved, more or less, by trial and error. Very roughly, one can say that the money supply should increase at the same rate as the output of the economy, or (for reasons that are not very well understood) slightly faster. It seems that a very slight rate of inflation is more desirable than zero inflation, because lowering interest rates is one of the main tools of the central bank and this tool becomes useless in the absence of inflation. Monetary policy at the month-to-month level is quite tricky, needless to say.

The fact that the *nominal* (or face) value of money need not coincide with its *real* value – because governments have the power to print money – introduces analytic difficulties. In particular, there are significant difficulties in attaching monetary value to produced goods, or to natural capital assets. Yet, despite the points noted above, the only practical answer is, as remarked at the beginning, that goods (and assets) must be valued by markets, taking care to allow (insofar as possible) for the fundamental difficulties.

Suppose we add up the total 'liquid' money assets of all economic units in a nation. And what determines the size of the basic *liquid* money supply (known as M1), consisting of cash, bank deposits and short-term credit (known as 'commercial paper') in relation to total national welfare? In fact, nobody – not even the Federal Reserve Bank – has a really good theoretical answer to this question. Yet this form of money is the lubricant of trade.

Without it, trade would revert to barter, and would virtually halt. Hence liquidity, *as such*, evidently has economic value. The Federal Reserve Bank can increase or decrease the current money supply (M1) by loosening or tightening bank credit and changing interest rates. (The detailed mechanisms do not concern us.) The impact on economic activity is fairly straightforward, albeit delayed by some months: increasing the liquid money supply is expansionary (and inflationary), while reducing the rate of expansion has the opposite effect, other factors remaining equal.

Evidently a link between the liquid money supply (M1) and wealth – and welfare – must exist. Yet the value of money and other financial assets possessed by agents surely cannot be equated to the value of stocks of goods already in existence, since the latter are already counted as part of the book value of firms and as part of the personal wealth of individuals, measured in terms of current (or constant) prices. There is a natural (if unknown) relationship in the real economy between aggregate liquidity (M1) and aggregate wealth and welfare. For purposes of this book, we must leave it there.

However, we think the arguments summarized above suffice to justify our assertion that money, as such, constitutes a major component of wealth, for both individuals and firms.

NOTES

1. Croesus used his wealth to conquer most of the nearby Greek cities in Asia Minor. Then he foolishly attacked the Persian empire and lost everything in 546 BC. But that is another story.
2. Incidentally, the Koran defines usury more precisely, and prohibits it more strictly than the Catholic Church, and one of the consequences is that Muslim countries have been slow to develop modern financial institutions.
3. Law's scheme, first adopted in France in 1716, was to back the bills (i.e. money) issued by his new 'Banque Générale' with future revenues from the Mississippi (and other) territories claimed by the French crown. The bills issued by law's bank – actually shares – were decreed to be acceptable for purposes of tax payments, which made them acceptable for other purposes as well. At first the scheme was wildly successful, and the rate of interest in France fell to 4.5%. Soon public enthusiasm encouraged the issuance of more and more shares, guaranteed by the king (in 1718) but backed by nonexistent revenues. The Mississippi bubble collapsed in 1720 and Law was discredited and persecuted. A very similar scheme was initiated in England, with notes (shares) backed by revenues from the South Sea company, which offered (1720) to take over the national debt by exchanging government annuities for shares. It, too, became wildly popular for a few months before collapsing (the South Sea Bubble), although the trading company was real and did have revenues.
4. This was a major factor in US political campaigns from 1888 through 1900, especially during the two presidential campaigns of Democrat William Jennings Bryan against Republican William McKinley. Bryan advocated 'free silver' or 'bimetallism', by which he meant that the US money supply should be backed by silver, as well as gold. The famous children's story *The Wizard of Oz* was originally a political satire, in which the yellow brick road represented gold, of course, and the wizard himself was supposed to be Mark Hanna, the industrialist and Republican 'éminence grise' behind William McKinley.

5. The history of the petroleum industry perfectly exemplifies this shift. In the year 1899 Standard Oil (the holding company) was recapitalized from a mere $10 million to $110 million, although it dominated the world oil business and owned substantial shares of 41 other companies that indirectly controlled scores of others. During the nine-year period 1893–1901 it paid out $250 million to its shareholders, representing several hundred percent returns *as dividends*, each year, even on the recapitalized stock (Yergin 1991, p. 99). Similarly, during the years 1903–05 the Royal Dutch Company paid annual dividends of 65%, 50% and 73% (in contrast to Shell, which paid only 5% and was consequently forced into the merger with Royal Dutch on unfavorable terms (ibid., p. 126).)

Appendix B: Balance equations; accounting relationships

B.1   DISCUSSION OF BALANCE EQUATIONS

In economics, commodities play almost as fundamental a role as firms and consumers, notwithstanding the importance of services. 'Commodity fetishism', despite its loud denunciation by Karl Marx, is indispensable to economics. The aim of all primary economic activity is to ensure material goods for consumption directed toward the satisfaction of human needs. Even in the modern service-oriented economy, most services are provided by material products. The role of stocks (commodities, goods, capital, money) is to satisfy both immediate and future consumption desires. As goods are physical, the laws of physics apply. The energy and mass conservation laws are rigorous bookkeepers. These balances are never violated.

Let n and k be the total number of agents and goods present in an economy, respectively. Let $X_{\alpha,i}(t)$ denote the quantity of stock of good I possessed by agent α. Money can be regarded as a special good, with the index $i = n$. Each agent's possessions (wealth) can be represented by its stock vector $X_{\alpha,i}(t) = (X_{\alpha,i}(t), i = 1,2, \ldots ,k)$.

Elements of the vector X are measured in natural units, so they are always non-negative. Tangible goods can be measured in mass (kilogram or kg) or exergy (kilojoule) units.[1] The conservation law of mass says that no mass can be created or destroyed. Mass can be transformed from one form to an other, or it can be transported from one place to another, but the total of inputs and outputs is always constant.[2] Second Law constraints are expressed in terms of exergy balances.

The stocks of goods usually can be measured in natural (mass or volume) units or in some equivalent economically relevant terms (e.g. pieces, bales, bushels, barrels, etc.). In the following we summarize the balance equations in a form which is simultaneously physically and economically valid. That is, the measuring unit will be physical (mass) and the implied actions are economic.

The Z-function (Chapter 4) normally depends on stocks of man-made goods (with certain exceptions, notably agriculture, forestry, fishing and mining). Its time-dependence is defined through the changes of stocks, which occur during trade, production or consumption processes. Nevertheless in production and consumption most transformations require labor and other material flows such as air and water, which are usually not stocked, but extracted, used and discarded immediately. Water may or may not be a free good; air is always a free good. Since the stocks of free goods are normally zero, they do not appear in the wealth function.

Services, including labor, do not appear directly in the wealth function (their stocks also being zero). However, those services that are used (or produced) in economic processes are nevertheless evaluated through changes in the wealth function.

We now introduce a production (transformation) vector T_i the elements of which are quantities of goods, services and/or 'free gifts' from nature used (converted, transformed) in a unit time ($T_i < 0$), or produced ($T_i > 0$). When all process inputs and outputs are taken into account the mass balance law holds in the form:

$$\sum T_i = 0 \qquad \text{(B.1)}$$

From this point on, the flow variable J (defined below) includes services and wastes as well as tangible material goods. Non-monetary stock changes satisfy a general balance equation. Money will be considered later. We can write:

$$\frac{dX_i}{dt} = J_i + T_i \qquad \text{(B.2)}$$

where X_i is the quantity of the ith stock of the economic agent, J_i is the net trade flow to/from other economic agents and T_i is the quantity of stock *transformed* during a production or consumption process, discussed in more detail later. In the case of services (from other agents or from nature) there is no possibility of storage (except for water) hence their stocks must always be zero. This implies that in case of services the production/use and the exchange are simultaneous. The flow of services J_s and the transformed quantity T_S always cancel each other.

$$\frac{dX_s}{dt} = J_s + T_s = 0 \qquad \text{(B.3)}$$

To describe the situation quantitatively, let y be the level of production (output/unit time) of the main product, that is:

$$T_p = y \qquad \text{(B.4)}$$

It is convenient to introduce a normalized production (transformation) vector u_i indexed over all inputs and outputs, including wastes, namely

$$u_i = \frac{T_i}{y} \tag{B.5}$$

where $u_p = 1$. The vector components corresponding to inputs are negative numbers, with absolute values equal to the mass of the input divided by the mass of the main product. Similarly, for by-products and waste products the vector elements are positive and equal to the mass of the by-product divided by the mass of the main product. In other words, u_i is the quantity of ith stock used for the production of a unit of final product (if $u_i < 0$) or the quantity of products and by-products, including wastes ($u_i(y) > 0$).

We can now describe the implications of mass balance for the main economic processes, as follows:

Trade of goods with another agent.   Conservation of mass implies that the total combined stock held by the two agents does not change, viz.

$$J_{\alpha\beta} = -J_{\beta\alpha} \tag{B.6}$$

This equation says that if a trade process only occurs between agents α and β, it does not modify the total stocks of the two agents, or the total stock of the economic system. Hence it follows that, for all i

$$\frac{d(X_{\alpha,i} + X_{\beta,i})}{dt} = 0 \tag{B.7}$$

Implicit trade with 'nature' is taken into account by adding an index. Trade with nature consists of raw materials extraction from farms, forests, mines, quarries, rivers, groundwater and the atmosphere itself, as well as waste disposal. The balance equation for the total stock is

$$\frac{d(X_{\alpha,i} + X_{n,i})}{dt} = 0 \tag{B.8}$$

where $X_{n,i}$ is the natural stock of the ith good.

Production of a good by a firm or production of labor by a consumer   This category includes production of goods from pre-existing stocks of materials, including fuels and (in the consumer's case) 'from' food, shelter and other necessities. In any transformation to another form the mass conservation law holds, provided all process inputs and all production-related wastes are taken into account. Production can be visualized as a transformation of an input vector to an output vector. The inputs include stocks of 'raw materials' or other goods; services from other agents and 'free gifts' from nature (air, water, sunlight, minerals, etc.). The outputs include the main product, by-products (saleable), and wastes (unsaleable by-products).

Production based in part on gifts of nature In reality there is practically no physical production that does not exploit some natural resource or service, at least indirectly. There are two categories, those which appear in the stock vector of the agent, and those that do not. For cases in the first category, where the resource is *exhaustible,* mass conservation implies a decrease in the stock of the remaining resources still in the ground, which matches the flow of materials extracted for processing. Mines, quarries and oil/gas wells are examples. A stock of underground resources is extracted, converted to a more useful form (by concentration, smelting or refining) and unusable waste materials are returned to the environment, often on-site. The conservation condition for the agent is:

$$\frac{dX_{\alpha,f}}{dt} = J_{\alpha n,f} \tag{B.9}$$

The complementary condition for the in-ground resource stock is

$$\frac{dX_{n,f}}{dt} = -J_{\alpha n,f} \tag{B.10}$$

In the second case, where the free gift of nature is not included in the agent's stock vector, there is no storage of inputs. Here it is useful to define two sub-categories. In the first, the free good is *embodied* in the product. In the second category it is not embodied, and is discarded almost immediately. The first case applies primarily to products of photosynthesis or animal husbandry, and a few chemicals. An agent (e.g. a farmer) cultivates plants that use free goods (water, carbon dioxide, sunlight) as direct inputs to production. But photosynthesis transforms the inputs (CO_2, H_2O) immediately to another form (e.g. CO_2 plus H_2O carbohydrate). Similarly, grazing animals convert grass plus oxygen to CO_2 and meat or milk. An example from the chemical industry is ammonia production, where nitrogen from the air reacts with hydrogen from natural gas at high temperatures and pressures. There are other examples (methanol, ethylene oxide, propylene oxide) where chemical feedstocks are partially oxidized, using atmospheric oxygen. Water treatment, now becoming a separate industry, is also an example.

Meanwhile the productive capital stocks of the farmer and forest-owner (land, animals) or the fisherman (boats), or the ammonia producer are essentially unchanged, except for wear. The conservation law also holds for 'trade' between agent α and the natural environment, indexed by n below:

$$\frac{dX_{\alpha,f}}{dt} = J_{\alpha n,f} + T_f = 0 \tag{B.11}$$

and

$$\frac{dX_{n,f}}{dt} = -J_{\alpha n} \tag{B.12}$$

The first of these equations reflects the fact that, because the stock of goods in agent α's inventory does not change there must be a compensating 'trade' (extraction or disposal) from or to the natural environment and/or a transformation from some other stock to account for it. (Usually it is one or the other but not both.)

The second equation indicates that there is a similar balance in the atmosphere (or, conceivably, in the lithosphere or in the hydrosphere). There are natural processes that regenerate atmospheric oxygen and nitrogen, but they are not linked to the economic process in question. Nevertheless, the economic outputs from these sectors (grain, meat, wood, fish, ammonia, clean water) do reflect flows to and from the environment of the 'free gift' type.

Free gifts from nature in the above examples will be incorporated into the production vector; thus

$$u_f = \frac{T_f}{y} \tag{B.13}$$

where u_f is the quantity of free good needed to produce one unit of final product, (if $u_f < 0$), or the quantity of $u_f < 0$ wastes freely disposed of (if $u_f > 0$).

The second sub-category of processes in which the agent's stocks do not change refers to production processes in which the gift of nature is not actually embodied in the product. In particular, air and water are important inputs to many other industrial processes, especially processes that involve combustion, washing, flotation, dilution, cooking, or chemicals. In reality, virtually all real economic transformation processes involve either production or consumption wastes. For these processes $dX_n/dt \neq 0$. There are significant waste flows in most industrial sectors as well as consumption wastes from final users.

Unfortunately, carbon dioxide inputs from industrial activity are *not* balanced, but accumulate in the atmosphere and the oceans, This is also true of nitrogen oxides and sulfur dioxide, as well as a number of halogenated compounds and heavy metals in potentially mobilizable (soluble) forms. These all result in unbalanced environmental stock changes that can also lead to uncontrolled changes in environmental cycles (Ayres 1997).

Total change of stocks (trade + production together). Summarizing, stock changes can be accounted for in terms of trade and production, plus exchanges with the environment. In the present form $J_{\alpha,\beta}$ describes ownership (property right) changes

$$\frac{dX_{\alpha,i}}{dt} = \sum_{\beta} J_{\alpha\beta,i} + yt_{a,i}(y) + J_{\alpha n,i} \tag{B.14}$$

(in the case of services $X_i = 0$ and $\dfrac{dX_i}{dt} = 0$)

The special case of money. In the case of money there are no producers except the government (central bank) and certain specialized institutions (banks) that are legally permitted to create money at a fixed rate (based on deposits) in the form of additional credit (see Chapter 7, Section 7.4). No other economic agent may create new money. So the agent can only get money from other agents, or he/she may give it to other agents,

$$\frac{dM_{\alpha}}{dt} = \sum_{\beta} J_{\alpha\beta,M} \tag{B.15}$$

where, because of the conservation of money in pair-wise exchange,

$$J_{\alpha\beta,M} = -J_{\beta\alpha,M} \tag{B.16}$$

In parallel to the goods case considered above, we divide the *money flow* term into two components. The first is the money payment for goods or services trade, that is, price times quantity $-p_{\alpha\beta,i}J_{\alpha\beta,i}$ comprising income from sales of goods or labor and payments for goods or services (wages are the price of labor.) The second component $I_{\alpha\beta}$ is a pure money flow, representing interest, taxes, royalties, rents or gifts. We ignore the possibility of currency exchange (money for money), since it is mainly a business of banks.

Then

$$J_{\alpha\beta,M} = -\sum_{i} p_{\alpha\beta,i}J_{\alpha\beta,i} + I_{\alpha\beta} \tag{B.17}$$

and the first term on the right-hand side is the money payment received for the quantity J_i of the ith good (or service) sold and transferred from α to β and p_i is the unit price of that good or service (including labor wages). The overall balance equation for money for the αth agent is

$$\frac{dM_\alpha}{dt} = \sum_\beta \left(-\sum_i p_{\alpha\beta,i} J_{\alpha\beta,i} + I_{\alpha\beta} \right) \tag{B.18}$$

In 'pure' economic processes, with no involuntary payments (e.g. taxes), and obeying the no-loss rule, we can set $I_{\alpha\beta} = 0$.

It is worthwhile to mention that the above equation, summed over all the agents in the whole economic system – if we exclude the actions of money-creating entities (banks) – provides a conservation law for money. Money, like energy, is neither created nor destroyed, in the short run, by normal economic processes. Hence

$$\sum_\alpha \frac{dM_\alpha}{dt} = \sum_\alpha \sum_\beta \left(-\sum_i p_{\alpha\beta,i} J_{\alpha\beta,i} + I_{\alpha\beta} \right) = 0. \tag{B.19}$$

In the longer time perspective, where banks are identified as specialized economic agents (consuming labor services and producing credit as a multiple of deposits), then there will be a source term on the right-hand side in place of the zero (Chapter 7).

NOTES

1. Exergy is the technical term for energy that is 'available', that is to do useful work. Energy is conserved in every process, whereas exergy is not. In fact, energy is destroyed in every process, as entropy increases.
2. Nuclear processes do result in an infinitesimal change of mass (as mass is converted into energy) but this can be ignored in practice.

Appendix C: Explicit representations of value and wealth functions and supply–demand curves

C.1   DERIVATION OF WEALTH FUNCTIONS

In this Appendix we consider some possible explicit wealth and value functions for three cases. The first case might be a pure trader of general goods, such as a country store, for which market decisions concerning the ith good in stock are independent of the quantities of the other goods it has in stock. In other words, for this particular trader the subjective value of the ith good does not depend on the quantities of the other goods held in stock.

The second case might also be a trader, except that it is the *marginal* contributions of stock changes to wealth that are assumed to be independent. We will see that the corresponding functional forms are different.

The third case is much more general. It applies to an agent such as a producer, whose stocks of goods are *inter-dependent*, either in value or quantity terms. This situation might arise because of the requirements of a production technology with either fixed (Leontiev-type) or variable input coefficients that are linked. Or it might apply to consumers, for whom the *combination* of stocks (e.g. books and shelf space, or dresses and shoes) is important. For this individual a change in one kind of stock changes the desired levels of other stocks, and hence his/her personal valuation of the combination.

There is a convenient restriction on the Z-function that can be applied equally to all cases. This arises from the assumption that doubling all stocks and money will double the wealth of the agent.[1] This is tantamount to requiring that Z should be a first-order homogeneous function of its arguments, known as the Euler condition. Note that this condition precludes the possibility of declining marginal value of wealth as a function of the arguments. Such a condition is also possible, of course, although we do not consider it here.

Case 1: Wealth functions for general goods traders

To begin, we consider an economic agent that buys and sells a variety of trade goods for money, but in small quantities so as to have a large selection on hand. Examples might be a pawnshop or a general store in a small town. The various goods in stock are needed only for future trade, so we can assume that the decisions to sell or to buy are defined by the expected price, the stock of money available and the current stock of the good. As the decisions to buy or sell are entirely based on monetary value it follows that the value of a good to the agent is not affected by the presence or absence of other goods.

Mathematically, the above condition means that

$$\frac{\partial v_i}{\partial X_k} = 0 \quad \text{if } i \neq k \tag{C.1}$$

In words, the marginal value of the ith good to the trader depends only on the quantity of that good, and the quantity of money available to the trader:

$$v_i = v_i(X_i, M) \tag{C.2}$$

The Euler condition for the wealth function (first order homogeneity) implies that the marginal value v_i of each good must be a homogeneous zeroth order function of the stock of that good and money. It follows that v_i depends on the *ratio* of stocks and money, viz.

$$v_i = v_i\left(\frac{X_i}{M}\right) \tag{C.3}$$

One can derive a useful relation for the homogeneous linear wealth function by differentiation, namely

$$X_i \frac{\partial v_i}{\partial X_i} = -\left(M + \sum_k v_k X_k\right) \frac{\partial \ln(w_M)}{\partial X_i} \tag{C.4}$$

Note that, as the left-hand side (l.h.s.) does not depend on X_k, the right-hand side (r.h.s.) must also not depend on X_k. Then, differentiating both sides with respect to X_k we obtain

$$\left(v_k + X_k \frac{\partial v_k}{\partial X_k}\right) \frac{\partial \ln(w_M)}{\partial X_i} + \left(M + \sum_l X_l v_l\right) \frac{\partial^2 \ln(w_M)}{\partial X_i \partial X_k} = 0 \tag{C.5}$$

After substituting the r.h.s. of *Equation C.4* into the l.h.s. of *Equation C.5* and rearranging terms, we obtain

$$\left(v_k + X_k \frac{\partial v_k}{\partial X_k}\right)X_i \frac{\partial v_i}{\partial X_i} + \left(M + \sum_l X_l v_l\right)^2 \frac{\partial^2 \ln(w_M)}{\partial X_i \partial X_k} = 0 \qquad (C.6)$$

Performing the same operations with indices k and i reversed we get a similar form,

$$\left(v_i + X_i \frac{\partial v_i}{\partial X_i}\right)X_k \frac{\partial v_k}{\partial X_k} + \left(M + \sum_l X_l v_l\right)^2 \frac{\partial^2 \ln(w_M)}{\partial X_k \partial X_i} = 0 \qquad (C.7)$$

The second terms in *Equations C.6* and *C.7* agree (because of Young's theorem), which implies that

$$\frac{X_i}{v_i} \frac{\partial v_i}{\partial X_i} = \frac{X_k}{v_k} \frac{\partial v_k}{\partial X_k} \qquad (C.8)$$

Since the l.h.s. is independent of X_k and the r.h.s. is independent of X_k it follows that both sides are equal to a constant:

$$\frac{X_i}{v_i} \frac{\partial v_i}{\partial X_i} = \gamma \qquad (C.9)$$

Inasmuch as v_i depends on $\frac{X_i}{M}$ we can multiply both sides by M, then

$$\frac{\partial \ln(v_i)}{\partial \left(\frac{X_i}{M}\right)} = \gamma \frac{M}{X_i} \qquad (C.10)$$

whence the general solution for marginal value takes the form:

$$v_i = c_i \left(\frac{M}{X_i}\right)^\gamma \qquad (C.11)$$

where γ is a parameter independent of i, and c_i is also a constant. All values depend on the relevant stock with the same exponent. Economic arguments forbid $\gamma < 0$, which would imply that the marginal value of goods would decrease with increasing money, and increase with increasing stocks.

Inserting *Equation C.11* back into *Equation C.10* we get

$$\frac{\partial \ln(w_M)}{\partial X_i} = -\frac{X_i M^{\gamma-1}}{X_i^{\gamma-1}\left(1 + \sum_k \left(\frac{M}{X_k}\right)^{\gamma-1}\right)} \tag{C.12}$$

Marginal value of money is obtained by integrating *Equation C.12*. The form of the integral depends on γ. The resulting wealth functions can be derived easily for two extreme values of the parameter γ.

The extreme case $\gamma = 0$ can be obtained from *Equation C.11*. The marginal value of the ith stock is constant, namely $v_i = c_v/c_0$. The marginal value of money in this case is also constant, $w_M = c_o$. The wealth function then takes the simple linear form

$$Z = \sum_i c_i X_i + c_0 M \tag{C.13}$$

Equation C.13 satisfies an important asymptotic condition; namely that the wealth of the agent does not vanish until all stocks go to zero.

However this functional form implies an agent that does not change its subjective valuation (or value-related decisions) even if its stocks change radically. In other words, the agent in this case does not adapt its buying and selling behavior to stock changes, which means the agent does not learn from experience. The trading agent buys if the price offered is less than the value it has established, and the value does not change, so the agent may continue buying even when it cannot sell. Conversely, this agent sells at any price higher than its initial subjective valuation, even when it cannot buy. Based on this argument, it would seem that this type of trading agent (i.e. one with a sum-type wealth function) must be very rare, if it exists at all. We therefore discard the additive choice of form.

The other extreme case is $\gamma = 1$. In this case

$$\frac{\partial \ln(w_M)}{\partial X_i} = -a_i X_i \tag{C.14}$$

The marginal value of money now takes the product form

$$w_M = a_0 M^{a_0-1} \prod_i^n X_i^{a_i} \tag{C.15}$$

where (to satisfy the Euler condition)

$$a_0 = 1 - \sum_i a_i \qquad \text{(C.16)}$$

and the Z-function becomes

$$Z = M^{a_0} \prod_i^n X_i^{a_i} \qquad \text{(C.17)}$$

This function is plotted in *Figure C.1* for M = 100 and $\alpha_0 = \alpha_1 = \dfrac{1}{2}$

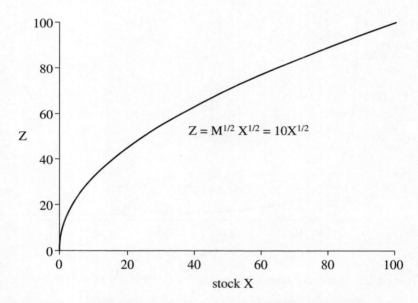

$$Z = M^{1/2} X^{1/2} = 10X^{1/2}$$

Figure C.1   Wealth as a function of stock. Z = X$^{1/2}$M$^{1/2}$ *for M = 100*

Equation C.17 leads to the following marginal value functions

$$w_M = a_0 \frac{Z}{M} \qquad \text{(C.18)}$$

and

$$v_i = \frac{a_i}{a_0} \frac{M}{X_i} \qquad \text{(C.19)}$$

Here, as before, the marginal subjective value of the ith good to the agent does not depend on the marginal value (or quantity) of the kth stock, in agreement with the original assumption. However, in this case the marginal subjective value of each stock tends to infinity as the quantity in store approaches zero, while, on the other hand, the marginal subjective value becomes infinitesimally small as the quantity of the good in stock becomes very large. The value of a stock is always positive, in this case, so there is no upper limit for the quantity of stocks that the trader might wish to own. (*Figure C.2* shows the value as a function of stock for the same case as *Figure C.1*).

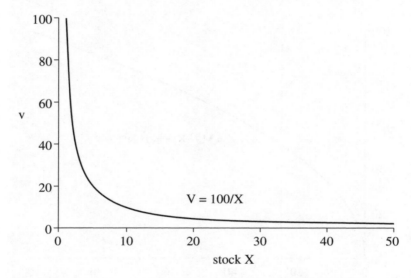

Figure C.2   Value as a function of stock. $v = (M/X)^{1/2}$ *for* $M = 100$

Evidently the agent's wealth in this case tends to zero when *any* of the stocks tends to zero. This sort of behavior is unrealistic for a pure trader. A trader who runs out of one good can always trade another good (assuming, as always, the existence of a market). The asymptotic dependence of wealth on stocks associated with *Equation C.17* is reasonable only for a specialized producer, or a consumer, all of whose inputs are essential and absolutely needed for the agent's economic survival. For instance, a baker must have flour, yeast and fuel for the oven; a painter must have paint, solvent and paintbrushes. If any of the essentials are missing, the agent cannot function.

The product type Z-function is obviously similar in form to the so-called Cobb-Douglas (C-D) production function that is so familiar to economists. There is a further similarity in that the product form of the arguments of the

C-D production function (capital, labor, materials) implies mutual substitutability. On the other hand, the functions have very different meanings and must not be confused.

For intermediate values of γ ($0 < \gamma < 1$), the parameter defined following *Equation C.12* it can be shown that

$$w_M = \left(1 + \sum_i c_i \left(\frac{X_i}{M}\right)^{1-\gamma}\right)^{\frac{\gamma}{1-\gamma}} \tag{C.20}$$

and the marginal value of the ith good is then

$$v_i = c_i \left(\frac{M}{X_i}\right)^{\gamma} \tag{C.21}$$

while the Z function becomes

$$Z = \left(M^{1-\gamma} + \sum_i c_i X_i^{1-\gamma}\right)^{\frac{1}{1-\gamma}} \tag{C.22}$$

which vanishes, as it should, only when all the stocks and money go to zero.

Case 2: Wealth functions when contributions to wealth are independent

For the next case we assume that the marginal contribution to wealth attributable to any one good does not depend on the quantity of other goods in stock. This condition would also apply to a trader or merchant, rather than to a producer or consumer. Here the separability condition comes into play in a different way and Z can be rewritten, as follows:

$$Z = \sum_i^n f_i(X_i, M) \tag{C.23}$$

Applying the homogeneous linearity (Euler) condition

$$Z = \sum_i^n X_i G_i(X_i, M) + c_0 M \tag{C.24}$$

where G_i is any zero[th] order homogeneous function of the arguments. For instance, one possible choice is

$$G_i = c_i \left(\frac{M}{X_i} \right)^{a_i} \tag{C.25}$$

where a_i and c_i are constants. This function has the necessary asymptotic property of vanishing when all the stocks go to zero. The marginal value of money to the agent is then given by

$$w_M = c_0 + \sum_i^n a_i c_i X_i^{1-a_i} M^{a_i-1} \tag{C.26}$$

and the marginal value of the ith good is

$$w_i = c_i (1 - a_i) \left(\frac{M}{X_i} \right)^{a_i} \tag{C.27}$$

Hence the marginal subjective value of the ith good, $v_i = \dfrac{w_i}{w_M}$, takes the form:

$$v_i = \frac{c_i(1 - a_i) M^{a_i} X_i^{-a_i}}{c_0 + \sum_k^n a_k c_k X_k^{1-a_k} M^{a_k-1}} \tag{C.28}$$

In this case, again, the wealth of the agent remains positive and non-zero as long as the agent has a non-zero stock of any good:

$$Z = \sum_i c_i M^{a_i} X_i^{1-a_i} + c_0 M \tag{C.29}$$

Another possible choice of form for G is a logarithmic function of the ratios of money and goods, viz.:

$$G_i = g_i \ln \left(c_i \frac{M}{X_i} \right) \tag{C.30}$$

where g_i and c_i are coefficients to be specified (in principle, by observation). The corresponding wealth function is:

$$Z = \sum_i g_i X_i \ln \frac{M c_i}{X_i} \tag{C.31}$$

In this case the marginal subjective value of money to the agent is given by

$$w_M = \frac{\sum_i^n g_i X_i}{M} \tag{C.32}$$

and the marginal subjective value of the ith good becomes

$$w_i = g_i + g_i \ln\left(c_i \frac{M}{X_i} \right) \tag{C.33}$$

so the marginal monetary value of the ith good ($v_i = w_i/w_M$) becomes:

$$v_i = \frac{M g_i}{\sum_k g_k X_k} \left[1 + g_i \ln\left(c_i \frac{M}{X_i} \right) \right] \tag{C.34}$$

Logarithmic dependence of values on quantities is consistent with many psycho-sensory relationships (e.g. apparent brightness vs. light intensity or

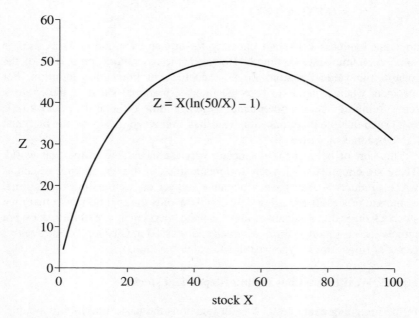

Figure C.3   Logarithmic wealth function: $Z = X [\ln (M/X) - 1]$ *for* $M = 50$

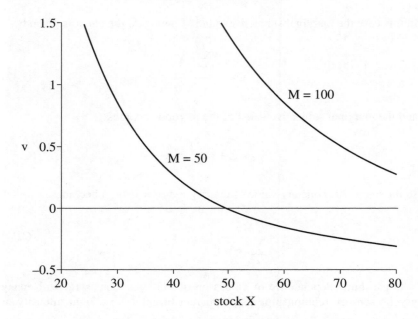

Figure C.4   Value as a function of stock X: *Logarithmic wealth function* v
= (M/X) ln (M/X)

apparent loudness vs. sound intensity for human eyes and ears) as well as some economic rules-of-thumb. However, the asymptotic properties of the logarithmic value functions do not accord with immediate intuition. For instance, when X_i tends to zero the internal marginal value of the ith stock tends to infinity. On the other hand, there is a finite value of the ith stock (c_iX_i = M) such that the marginal value vanishes, and when ($c_iX_i > M$) the marginal value becomes negative.

This sort of behavior is not necessarily inconceivable in the real world. There are circumstances where too much stock of the wrong kind – such as waste products – may have a negative impact on wealth. In most normal economic processes the agent deliberately avoids the regime where marginal stock changes have negative value, so the theoretical possibility does not violate the no-loss rule for quasi-cycles, discussed in Chapter 2). See *Figures C.3, C.4* (logarithmic-type wealth and value functions).

Case 3: Wealth functions for interdependent stocks

In the foregoing examples of wealth functions the basic simplifying assumptions were independence, or separability. However, by far the most realistic case is the one where stocks (and their values) are interdependent. To cite a simple example: an additional unit of gasoline has no marginal subjective value for an agent (or consumer) who does not own a car while, conversely, the value of an extra car to the agent is strongly dependent on the availability of gasoline.

To explore this case we now introduce a functional form that is consistent with interdependence of the above kind, a version of the so-called linear-exponential or LINEX function:

$$Z = X_1^{a_1} X_2^{a_2} M^{a_0} \exp\left(\frac{-X_1}{X_2}\right) = Z_0 \exp\left(-c\frac{X_1}{X_2}\right) \tag{C.35}$$

where X_1 stands for the stock of gasoline and X_2 stands for the stock of motor vehicles. The first order homogeneity (Euler) condition requires that $a_1 + a_2 + a_0 = 1$, and c is a constant. In *Figure C.5* wealth is plotted as a function of gasoline stock X_1 for fixed values of the other parameters.

The peak in the graph represents the optimum combination of the two stocks. The marginal subjective value of money in this example is

$$w_M = \frac{Z_0}{a_0 M} \tag{C.36}$$

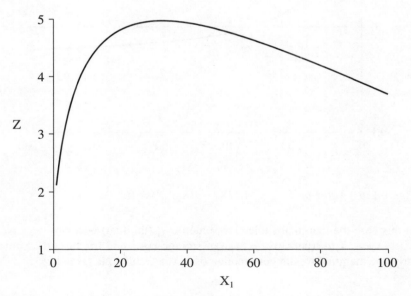

Figure C.5   *LINEX wealth function:* $Z = (X_1 X_2 M)^{1/3} \exp(-X_2/X_1)$, $X_2 = 10$, $M = 10$

which depends on the stocks of motor fuel and vehicles (or any other case of mutual dependence). The marginal contribution of the motor fuel to the agent's wealth function is

$$w_1 = \left(\frac{a_1}{X_1} - \frac{c}{X_2} \right)(X_1^{a_1} X_2^{a_2} M^{a_0})$$ (C.37)

while the marginal subjective value of the stock of fuel is

$$v_1 = \frac{M}{a_0} \left(\frac{a_1}{X_1} - \frac{c}{X_2} \right)$$ (C.38)

This equation is plotted in *Figure C.6*.

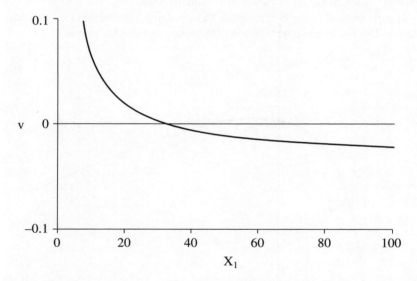

Figure C.6 *Value of* X_1. $v = M(1/X_1 - 3/X_2)$, $X_2 = 100$, $M = 10$

In this case the marginal subjective value of X_1 (the fuel) becomes negative when $X_2 > cX_1/a_1$ (perhaps because of storage costs and fire hazard). Similarly, the marginal wealth contribution of X_2 (the vehicle) is given by:

$$w_2 = \left(\frac{a_2}{X_2} + c \frac{X_1}{X_2^2} \right)(X_1^{a_1} X_2^{a_2} M^{a_0})$$ (C.39)

and the marginal subjective value of v_2 is

$$v_2 = \frac{M}{a_0}\left(\frac{a_2}{X_2} + c\,\frac{X_1}{X_2^2} \right) \tag{C.40}$$

Obviously the LINEX function can be extended easily to more complex examples.

C.2 WEALTH OF CONSUMERS

In our paradigm purchase decisions of goods by consumers are economic activities (comparison, bargaining) subject to the AAL rule. The subjective value of a good must exceed the price to justify such a purchase. The subjective (internal) value of a consumption good must depend on the quantities of other goods and money held by the consumer. On the other hand, consumption *per se* is a non-economic process. Consumption of purchased services, or of goods already in stock, contributes to the happiness or satisfaction of *H. Sapiens*. But it is a wealth loss for *H. Economicus* in the sense that Z, being a function of stocks of goods and money, must decrease. The question now arises: is there a way to accommodate the activity of consumption of purchased services within the AAL framework?

One rather artificial possible approach would be to regard the service as received *as if* it were a good, even though services cannot be stocked. This approach, in effect, treats a service as a non-material good with an extremely short lifetime. On the other hand, it would force us to think in terms of Z-functions changing from moment to moment, as the service is purchased, and then subsequently consumed.

Another approach is to recognize explicitly that the subjective value of services (and consumption in general) arises from changes in the state of the consumer that would – in a much more elaborate formulation – be included within the Z-function as *attributes*. For example, a haircut changes the appearance of the consumer. The consumer values the improved appearance more than the cost of the service. The improvement lasts for some time, whence the Z-function, if it reflected attributes, would not immediately decline by the amount of the money cost, but would increase by some 'attractiveness' factor.

To take another example, suppose the consumer buys a good restaurant meal and attends a theatrical performance. The immediate expenditure is presumably less than the subjective value received by the consumer. This can be both in the form of immediate enjoyment, as well as adding to a stock of pleasant memories. Other examples would be analyzed in a similar way.

However, for the purposes of this Appendix, and this book, the additional complexity that would be required to redefine Z-functions to include such non-material attributes is not justified by the additional explanatory power it would provide at this stage. We merely argue that a more elaborate future version of our wealth theory should be able to explain consumer behavior in terms of changes in attributes as well as physical stocks of material goods and money.

C.3   SUPPLY-DEMAND CURVES

In principle, the exchange process is simple: prices are announced by the 'market' and agents choose how much they wish to buy of each good (demand) and how much they wish to sell (supply). If the market-wide demand for each good meets the market-wide supply of each good a Walrasian equilibrium exchange is said to be achieved. If aggregate demand is not equal to supply at the published price, then the price is too high and it must fall. Conversely, if the aggregate demand exceeds the supply at the published price, the price must be too low and will rise.

The neoclassical exchange process is thus price-mediated in the sense that there is a price that is 'given' by the market and which the agents or actors in the market, effectively, treat parametrically. Note that individual agents must decide to buy or sell (or not) on the basis of the price of the moment, or on the basis of expectations of future price movements. If the market price falls, those who bought earlier have overpaid and *vice versa*. If the price rises when the agent held off from buying in the expectation of a fall, that agent is either frozen out or must buy at a higher-than-expected price.

For convenience, we can now imagine that the prices of offers and sales completed are published by an all-knowing demon (not to be confused with a real auctioneer). The demon – which somewhat resembles the NASDAQ (formerly 'over the counter') stock exchange – announces the prices at which potential buyers are willing to buy and sellers are willing to sell various quantities of the good. The agent may make a counter-offer to buy at a lower price. The demon announces this new offer. Or the agent may select among the possibilities already on offer that one which yields the highest wealth increase (bearing in mind that the agent has an internal valuation for that good higher than the price of the offer.) After the trade is completed the demon announces the details of the latest trade (quantity and price).

Setting aside the reaction of the market to the trade, let us consider an agent with one kind of good, and money, with initial stocks as follows: $X = 50$, $M = 100$. Suppose the agent has a wealth function of the product form:

$$Z = \sqrt{XM} \qquad (C.41)$$

Suppose also that the initial wealth of the agent is 70, and the marginal value of the good is:

$$Z = \sqrt{50 \times 100} = \sqrt{5000} \cong 71$$

Solving, we get $v = M/X = 2$

Suppose that the demon announces a market price $p = 3$. The agent sees that it is not advantageous to buy, as its subjective marginal value is less than the price. The AAL rule therefore forbids buying.

Suppose, now, that the demon announces a lower price $p = 2$. The agent will be indifferent, as its marginal subjective value is just equal to the proposed price. An exchange does not change the agent's wealth. The driving force in this case is zero and a real trade will not occur in this case either.

Finally, suppose the demon announces a still lower price $p = 1$. The agent now finds it is advantageous to buy, as its marginal subjective value is higher than the price. The question is how much should it buy? The agent seeks the trade that will result in the maximum increase in its wealth. In this case, it can be calculated. After buying a quantity q, the agent's new wealth would be

$$Z = \sqrt{(X + q)(M - pq)} = \sqrt{(X + q)(M - q)} \qquad (C.42)$$

where $p = 1$. This solution is plotted in *Figure C.7*.

To maximize its gain, the agent – now in a mood to buy – needs to decide on the quantity to be purchased. To find this point analytically, the agent (and we) can differentiate *Equation C.42*. It is easy to show that the agent's wealth is maximized when $q = M/2p$. In our example, the maximizing agent buys $q = 25$ units at price $p = 1$. If the price decreases further the demand increases, as $q \to M/2p$. Demand in this case approaches infinity as p approaches zero. (This is a consequence of our choice of the product form of the wealth function.) In the case of a LINEX wealth function there would be a finite demand at zero price, which is more realistic, as illustrated in *Figure C.8*.

For the case of logarithmic wealth function we selected the initial stocks as in the previous case, namely $X = 50$, $M = 100$. The constants were selected such that the initial marginal value was 2 and $p = 2$ was the equilibrium price, where supply equals demand and the value-price difference vanishes. The q (supply-demand) is the solution of the marginal value = price equation, as follows:

$$p = \frac{M + pq}{X - q} \ln\left(c \frac{M + pq}{X - q} \right) \qquad (C.43)$$

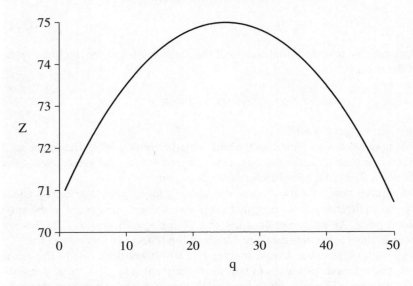

Figure C.7   Wealth as a function of trade. $M = 100$, $X = 50$, $p = 1$

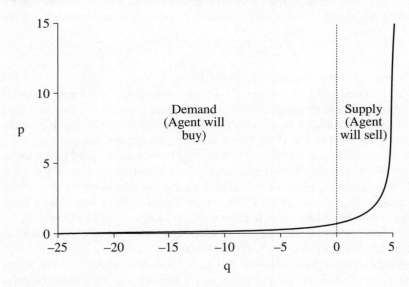

Figure C.8   Quantity to sell or buy, LINEX function

In *Figure C.9* the price is plotted as a function of propensity to sell (supply) vs. propensity to buy (demand) on the part of the agent. Positive values of q reflect supply, while negative values of q reflect demand.

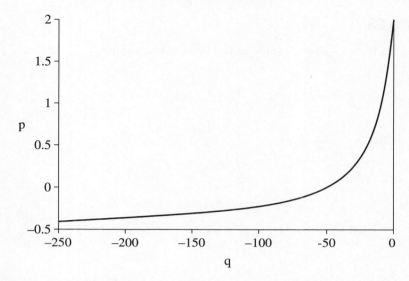

Figure C.9   Quantity to sell or buy logarithmic wealth function with M =
           100, X = 50

In this simple model, every agent is both a potential buyer and a potential
seller, depending on the price offered. In the logarithmic example we plotted
the supply-price curve for increasing prices. As the initial stock was 50 units,
we might assume that all will be sold, suggesting a maximum supply of 50.
That is not what happens. In this case there is a maximum supply (9.2) at
price p = 12. (This occurs off scale in *Figure C.9* and hence does not appear).
If we increase the price further the supply will actually decrease. This is due
to the non-linear dependence of value on money in the logarithmic function.

The shape of the curve far from equilibrium is essentially a mathematical
artifact that need not have an economic interpretation. (However a possible
interpretation is that beyond this point a higher price brings the seller so
much more money that the agent chooses not to work so hard. It might, for
instance reflect a tradeoff between money and leisure time.)

The other peculiar feature of the logarithmic case illustrated here is that
demand remains finite as price tends to zero, *and remains finite even for
negative prices*. This is consistent with our comments elsewhere that demand
for all goods need not satiate, whereas for some goods too much is actually
harmful.

NOTES

1. This is a consequence of constant returns to scale (Euler's condition).

Appendix D: Properties of the matrix L

If J_i is the flow of goods, then an element L_{ik} of the matrix L is the quantity of the ith good bought (or sold) due to the kth value-price difference. If J is interpreted as a demand vector, whose elements are demands for the ith good, then L_{ik} expresses the purchaser's willingness to buy/sell the ith good due to the kth value-price difference. A corresponding equation expresses the seller's willingness to sell. The vector J is common to the buyer and seller (except for the sign) and the agreed prices are also common to the two of them. However, the L matrices for the buyer and seller must be different if their subjective (internal) values are different. Thus L reflects the link between subjective values and demand (from the buyer's point of view) or between subjective values and supply (from the seller's perspective).

In brief, L can be regarded as a generalization of the conventional scalar demand function of microeconomics. For simplicity we call it the *demand* matrix hereafter. (The demand matrix is introduced here only for the case of trade. However, we generalize it later to account for production and consumption.)

The off-diagonal elements of the L-matrix are 'cross-effects' due to competition or substitution interdependencies with other goods. For instance, the price of oil and the price of gas are strongly correlated because of substitution possibilities between them. The diagonal matrix elements reflect demand that would occur in the absence of any such substitution possibilities. In other words, diagonal elements represent quantities of goods demanded depending only on the value-price difference for that good.

The diagonal elements are very nearly equivalent to the conventional demand functions for goods (or services), with one important exception: flows are quantities per unit time, while demand in the conventional picture (which appears in every elementary economics textbook) is simply the quantity purchased. There is another important difference, at least with respect to the textbook definition: In textbooks the marginal demand function for a 'good' is always positive, for any positive price. As mentioned already in Chapter 2, we argue that demand can saturate, goods can become bads (in excess quantities) and shadow prices can become negative.[1] These facts have important consequences in regard to the theoretical treatment of environmental services, toxic trace elements, pollution and wastes, for instance.

It is important to recognize that the magnitude of the matrix element L_{ik} depends on the situation, that is, with whom and how an agent exchanges goods. This means that the demand matrix L of an agent embodies strategies for interaction (e.g. bargaining) with other agents. It is modified by learned behavior. From a third-party standpoint these matrix elements can only be determined empirically, by observation or experiment. But, again, we emphasize that the elements of L may also depend on other factors.

D.1   SOCIAL CHOICE

The demand relationship derived above can also be generalized to an aggregation of buyers and sellers, that is, a *market*. In that case, assuming sellers and buyers to be distinct, we can interpret J as the (net) demand vector, v being the value vector of goods evaluated by the consumer(s), and p becomes the market price vector.

In pair-wise exchanges between bargaining agents (price makers) indexed by α and β, the above relation is sufficient to define the price at which the good is bought by one agent and sold by the other. The mass conservation rule for material stocks requires that mass is neither created nor destroyed. It follows that

$$J_{\alpha\beta,i} + J_{\beta\alpha,i} = 0 \tag{D.1}$$

Substituting the defining relationships, we can solve for p_i, the price at which the exchange takes place in terms of diagonal matrix elements

$$p_i = \frac{L_{\alpha,ii} v_{\alpha,i} + L_{\beta,ii} v_{\beta,i}}{L_{\alpha,ii} + L_{\beta,ii}} \tag{D.2}$$

or

$$p_i = \frac{v_{\alpha,i} + x v_{\beta,i}}{1 + x} \tag{D.3}$$

where x is the ratio of the diagonal matrix elements for α and β, viz.

$$x = \frac{L_{\alpha,ii}}{L_{\beta,ii}} \tag{D.4}$$

The agreed price is in the interval $v_{\alpha,i} \le p_i \le v_{\beta,i}$. The parameter x defines the price in this interval. In *Figure D.1* we have plotted the price for the case

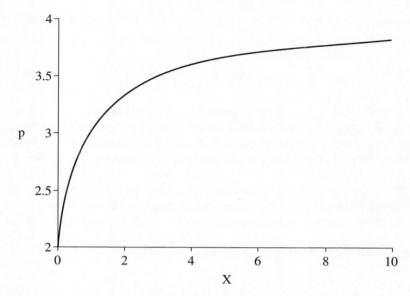

Figure D.1   Price agreement $p = f(x)$ *for* $v(\alpha, i) = 2$ *and* $v(\beta, i) = 4$

when $v_{\alpha,i} = 2$ and $v_{\beta,i} = 4$. The expected unit gain (i.e. the 'driving force') for agent α is

$$F_{\alpha,i} = v_{\alpha,i} - p = (v_{\alpha,i} - v_{\beta,i})\left(\frac{1}{1+x}\right) \tag{D.5}$$

The driving force for this agent approaches zero as x becomes very large. On the other hand, the driving force for the other agent β approaches a maximum as x tends to infinity.

$$F_{\beta,i} = v_{\beta,i} - p = -(v_{\alpha,i} - v_{\beta,i})\left(\frac{x}{1+x}\right) \tag{D.6}$$

Note that, in equilibrium, the driving force vanishes (because $p = v$). We discuss the equilibrium case briefly later. Substituting back into the expression for trade flow for agent α, we get

$$J_{\alpha,i} = \frac{L_{\alpha,ii}L_{\beta,ii}}{L_{\alpha,ii} + L_{\beta,ii}}(v_{\alpha,i} - v_{\beta,i}) = L_{\beta,ii}\left(\frac{x}{1+x}\right)(v_{\alpha,i} - v_{\beta,i}) \tag{D.7}$$

Similarly, the general expression for change of wealth function for agent α becomes

$$\frac{dZ_\alpha}{dt} = (v_{\alpha,i} - p_i)J_{\alpha,i} = \frac{xL_{\beta,ii}}{(1+x)^2}(v_{\alpha,i} - v_{\beta,i})^2 \tag{D.8}$$

It is now possible to treat the negotiation as a game and consider the 'best' strategy for agent α. This means choosing a value for the matrix element $L_{\alpha,i}$ assuming that $L_{\beta,i}$ (i.e. the propensity of agent β to sell good *i*) is known by agent α. There are at least three possible criteria for making this choice, viz.

To maximize the flow of goods implies maximizing $L_{\alpha,ii}$ whence $x \to$.
To maximize the unit gain implies minimizing $L_{\alpha,ii}$ whence $x \to 0$.
To maximize the total gain implies choosing $L_{\alpha,ii} = L_{\beta,ii}$ whence $x \to 1$.

Away from equilibrium, agent β also gains by the transaction by an amount.

$$\frac{dZ_\beta}{dt} = (v_{\beta,i} - p)J_{\alpha,i} = L_{\beta,ii}(v_{\alpha,i} - v_{\beta,i})^2\left[\frac{x^2}{(1+x)^2}\right] \tag{D.9}$$

Notice, that the two wealth increases are not the same. For agent β the optimal choice of agent α would be $x \to \infty$. In that case, all the wealth increase accrues to agent β. Of course agent α would prefer the opposite choice.

Assuming they both evaluate and negotiate, they may seek the joint optimal solution, namely the one that maximizes the increase of their *combined* wealth. This solution can also be calculated explicitly. For the optimal solution, $x = 1$, whence $L_{\alpha,ii} = L_{\beta,ii} = L$.

$$\frac{dZ}{dt} = \frac{(v_{\alpha,i} - v_{\beta,i})^2}{4} \tag{D.10}$$

Evidently the ratio *x* is a (partial) measure of the difference between the *L*-matrix elements of agents α and β. The maximum absolute increase in total wealth from trading occurs when the two matrices are equal, namely when *x* = 1 (i.e. $L_\alpha = L_\beta$). Remember the example of Jack and Jill, in Chapter 2, who maximized their total gain by sharing the trading gains equally.

Evidently, if the possibility for trade is restricted to a few agents only, then the two (or few) agents share a mutual incentive to modify their *L*-matrices so as to make them converge. Convergence can be interpreted as a learning process, such that the agents know more and more, over time, about how the other agent will respond in a given situation. It is quite plausible that cooperative behavior is essentially the end-result of this process of convergence.

If there are many agents in a market the mechanism for convergence is not so clear. Each agent has an incentive to increase its **L** by maximizing sales volume, in order to increase revenues. But it has a contrary incentive to decrease **L**, by raising prices, to increase the profit on each unit transaction. Real merchants face this dilemma all the time. The overall profit maximizing solution depends on local circumstances, including the strategies of other agents.

In reality L_β is not known by agent α or vice versa, so neither can simply 'select' the convergent social optimum $x = 1$. As it happens, however, there are simple rules by which an agent can accelerate this convergence process. Briefly, from the perspective of agent α:

(i) $\delta L > 0$ has the effect of decreasing the wealth increase (profit) per unit of stock exchanged, while increasing the flow (J). This results from a lower agreed price for selling and a higher price for buying, so the other agent will be willing to trade more.

(ii) $\delta L < 0$ has the opposite effect, namely a higher price for selling, and a lower price for buying, resulting in less trade.

For agent β the rules are the same. Obviously changes in either direction can go too far, since neither agent has any a priori way of knowing where the social optimum lies. However the tendency for the buyers and sellers to converge towards each other in the long run is clear.

Reality is complex and changing. Thus the true optimum value for **L** may change from transaction to transaction. Needless to say, in a multi-agent situation all the other agents learn also, from the same algorithm. As a rule, however, there is no time to find the true optimal **L**, even if such a solution exists in a multi-agent situation. A 'reasonable' value of **L** is the normal target.

NOTES

1. As noted in *Chapter 2*, many environmental services are normally provided at an optimum level, determined by co-evolution, which means that both deficit and excess are possible. For instance, our bodies need trace amounts of a number of metals such as copper, nickel, chromium, selenium and zinc, whereas too much of any is poisonous. Farmers may have too little water, in which case they will pay to get more, but too much water is a flood and farmers willingly pay to have the excess drained. Too little heat means crops (or ourselves) may freeze; but too much heat can be just as lethal to crops or people. A few glasses of wine can be good for you; too much alcohol can kill. Too little food is starvation; too much leads to obesity and early death.

References

Abramovitz, Moses (1956), 'Resources and output trends in the United States since 1870', *American Economic Review*, 46.

Albert, M., and R. Hahnel (1990), *A Quiet Revolution in Welfare Economics*, Princeton NJ: Princeton University Press.

Alchian, A. (1963), 'Reliability of progress curves in airframe production', *Econometrica* **31**, 679–93.

Alchian, A. A. (1950), 'Uncertainty, evolution and economic theory', *Journal of Political Economy* **58**, 211–22.

Andersen, Esben Sloth (1996), *Evolutionary Economics: Post Schumpeterian Contributions*, paperback edn, London: Pinter.

Argote, Linda, and Dennis Epple (1990), 'Learning curves and manufacturing', *Science* **247**, 920–4.

Arrow, Kenneth J. (1962), 'The economic implications of learning by doing', *Review of Economic Studies* **29**, 155–73.

Arrow, Kenneth J., and Gerard Debreu (1954), 'Existence of an equilibrium for a competitive economy', *Econometrica* **22** (3).

Arrow, Kenneth J., and Michael D. Intriligator (eds) (1981), *Handbook of Mathematical Economics, Book I*, vol 1, *Handbooks in Economics*, Amsterdam: North Holland.

Arthur, W. Brian (1983), *Competing Technologies and Lock-in by Historical Small Events: The Dynamics of Allocation under Increasing Returns*, Stanford University Committee for Economic Policy Research research paper 43, Palo Alto, CA.

Arthur, W. Brian (1988a), 'Competing technologies: An overview, in Giovanni Dosi, Christopher Freeman, Richard Nelson, Gerald Silverberg and Luc Soete (eds), *Technical Change and Economic Theory*, London: Pinter.

Arthur, W. Brian (1988b), 'Self-reinforcing mechanisms in economics', in Philip W. Anderson, Kenneth J. Arrow and David Pines (eds), *The Economy as an Evolving Complex System*, Redwood City CA: Addison-Wesley Publishing Company.

Arthur, W. Brian, Yu M. Ermoliev, and Yu M. Kaniovski (1983), 'The generalized urn problem and its applications', *Kibernetika* 1, 49–56.

Arthur, W. Brian, Yu M. Ermoliev, and Yu M. Kaniovski (1987a), *Nonlinear Urn Processes: Asymptotic Behavior and Applications*, International Institute for Applied Systems Analysis working paper WP-87-85 Laxenburg, Austria.

Arthur, W. Brian, Yu M. Ermoliev, and Yu M. Kaniovski (1987b), 'Path-dependent processes and the emergence of macrostructure', *European Journal of Operations Research*, 30, 294–303.

Axelrod, Robert (1984), *The Evolution of Cooperation*, New York: Basic Books.

Axelrod, Robert (1987), 'The evolution of strategies in the iterated prisoner's dilemma', in Davis, L. *Genetic Algorithms and Simulated Annealing*, Los Altos CA: Morgan Kaufmann.

Axelrod, Robert, and D. Dion (1988), 'The further evolution of cooperation', *Science* **242**, 1385–90.

Axtell, Robert, and Joshua Epstein (1996), *Growing Artificial Societies: Social Science from the Bottom Up*, Cambridge MA: MIT Press/Brookings Institution.

Axtell, Robert, Joshua Epstein, and Peyton Young (1999), 'The emergence of economic classes in an agent-based bargaining model'. In Steven N. Durlauf and Peyton Young (eds), *Social Dynamics*, New York: Oxford University Press/Sante Fe Institute.

Ayres, Clarence E. (1944), *The Theory of Economic Progress*, Chapel Hill, NC: The University of North Carolina Press.

Ayres, Robert U. (1978), *Resources, Environment and Economics: Applications of the Materials/Energy Balance Principle*, New York: John Wiley and Sons.

Ayres, Robert U. (1989), 'Technological transformations and long waves', *Journal of Technological Forecasting and Social Change* **36** (3).

Ayres, Robert U. (1994), *Information, Entropy and Progress*, New York: American Institute of Physics.

Ayres, Robert U. (1997), 'Integrated assessment of the grand nutrient cycles', *Environmental Modeling and Assessment* (2), 107–28.

Ayres, Robert U. (1998), 'Eco-thermodynamics: Economics and the second law', *Ecological Economics* **26**, 189–209.

Ayres, Robert U., Ralph C. d'Arge, and Allen V. Kneese (1970), *Aspects of Environmental Economics: A Materials Balance-General Equilibrium Approach*, Baltimore MD: Johns Hopkins University Press.

Ayres, Robert U., and Allen V. Kneese (1969), 'Production, consumption and externalities', *American Economic Review*, **59**, 282–97.

Ayres, Robert U., and Katalin Martinás (1992), 'Learning from experience and the life cycle: Some analytic implications', *Technovation* **12** (7), 465–86.

Ayres, Robert U., and Katalin Martinás (1996), 'Wealth accumulation and economic progress', *Evolutionary Economics* **6** (4).

Ayres, Robert U., and Indira Nair (1984), 'Thermodynamics and economics', *Physics Today* **37**, 62–71.

Baloff, N. (1966), 'The learning curve: Some controversial issues', *Journal of Industrial Economics* **14**, 275–82.

Barnett, Harold J. and Chandler Morse (1962), *Scarcity and Growth: The Economics of Resource Scarcity*, Baltimore MD: Johns Hopkins University Press.

Becker, Gary S. (1968), 'Crime and punishment: An economic approach', *Journal of Political Economy* **76** (March/April).

Becker, Gary S. (1973), 'A theory of marriage: Part I', *Journal of Political Economy* **82**, (July/August).

Binswanger, Hans P. and Vernon Ruttan (1978), *Induced Innovation: Technology, Institutions and Development*, Baltimore MD: Johns Hopkins University Press.

Boadway, R. (1974), 'The welfare foundations of cost-benefit analysis', *The Economic Journal* **84** (December), 926–39.

Boulding, Kenneth E. (1966), 'Environmental quality in a growing economy', In *Essays from the Sixth RFF Forum*, edited by H. Jarrett. Baltimore MD: Johns Hopkins University Press.

Boulding, Kenneth E. (1981), *Evolutionary Economics*, Beverly Hills CA: Sage Publications.

Bowles, Samuel (2004), *Microeconomics: Behavior, Institutions and Evolution*, in Colin Camerer and Ernst Fehr (eds), *Roundtable Series in Behavioral Economics*, Princeton NJ: Princeton University Press.

Bowles, Samuel and Herbert Gintis (2000), 'Walrasian Economics in Retrospect', *Quarterly Journal of Economics* **115** (4), 1411–39.

Brekke, K. and Richard Howarth (2000), 'The social contingency of wants', *Land Economics* **76** (November), 493–503.

Bródy, Andras, Katalin Martinás, and Konstantin Sajo (1985), 'Essay on macroeconomics', *Acta Oec.*, **36**, 305.

Bromley, D. W. (1990), 'The ideology of efficiency: Searching for a theory of policy analysis', *Journal of Environmental and Economics and Management* **19** (1), 86–107.

Candeal, Juan Carlos, Carlos Herves and Esteban Indurain (1998), Some results on representation and extension of preferences', *Journal of Mathematical Economics* **29**, 75–81.

Cartelier, J. (1990), *La formation des grandeurs économiques*, Paris: Presses Universitaires de France.

Cassel, Gustav (1932), *The Theory of Social Economy*, New York: Harcourt-Brace.

Chipman, J. and J. A. Moore (1976), 'Why an increase in GNP need not imply an improvement in potential welfare', *Kyklos*, **29** (3), 391–418.

Chipman, J. and J. A. Moore (1978), 'The new welfare economics 1939–1974', *International Economic Review* **19** (3), 547–84.

Clark, John, Christopher Freeman and Luc Soete (1983), 'Long waves, inventions and innovations', in Christopher Freeman (ed.), *Long Waves in World Economic Development*, London: Butterworths.

Cleveland, Cutler J. and Matthias Ruth (1997), 'When, where and by how much do biophysical limits constrain the economic process? A survey of Nicholas Georgescu-Roegen's contribution to ecological economics', *Ecological Economics* **22** (3), 203–24.

Coase, Ronald H. (1960), 'The problem of social costs', *Journal of Law and Economics* **3**:1–44.

Cobb, Clifford W. and John B. Jr. Cobb. (1994), *The Green National Product: A Proposed Index of Sustainable Economic Welfare*, Lanham MD: University Press of America.

Costanza, Robert (1991), 'Assuring sustainability of ecological economic systems', in Robert Costanza (ed.), *Ecological Economics: The Science and Management of Sustainability*, New York: Columbia University Press.

Daly, Herman E. (1985), 'The circular flow of exchange value and the linear throughput of matter-energy', *Review of Social Economics*, **43**, 279–97.

Daly, Herman E. (1992a), 'Allocation, distribution, and scale: towards an economics that is efficient, just', *Ecological Economics* **6**, 183–93.

Daly, Herman E. (1992b), 'Is the entropy law relevant to the economics of natural resource scarcity? Yes, of course it is!', *Journal of Environmental Economics and Management* **23**, 91–5.

Daly, Herman E. and John Cobb (1989), *For the Common Good*, Boston MA: Beacon Press.

Dasgupta, Partha and Geoffrey Heal (1974), 'The optimal depletion of exhaustible resources', paper read at Symposium on the Economics of Exhaustible Resources.

David, Paul A. (1975), *Technical Choice, Innovation and Economic Growth*, London: Cambridge University Press.

David, Paul A. (1985), 'CLIO and the economics of QWERTY', *American Economic Review* (*Papers and Proceedings*) **75**, 332–7.

David, Paul A. (1988a), *The Future of Path-Dependent Equilibrium Economics*, Stanford University Center for Economic Policy Research technical paper 155, August, Stanford, CA.

David, Paul A. (1988b), *Path-dependence: Putting the Past into the Future of Economics*, Stanford Institute for Mathematical Studies in the Social Sciences technical report 533, November, Stanford, CA.

Day, Richard H. (1984), 'Disequilibrium economic dynamics: A post-Schumpeterian contribution', *Journal of Economic Behavior and Organization* **5** (1), 57–76.

Day, Richard H. (1987), 'The general theory of disequilibrium economics and of economic evolution', in D. Batten, J. Casti and B. Johansson (eds), *Economic Evolution and Structural Change*, New York: Springer-Verlag.

Day, Richard H. and Gunnar Eliason (eds) (1986), *The Dynamics of Market Economies*, Amsterdam: North Holland.

Day, Richard H. and Jean-Luc Walter (1989), 'Economic growth in the very long run', in William A. Barnett, Ernst R. Berndt and Halbert White (eds), *Economic Complexity, Sunspots, Bubbles and Non-Linearity*, Cambridge, UK: Cambridge University Press.

Debreu, G. (1959), *Theory of Value*, New York: John Wiley.

Dorfman, Robert, Paul A. Samuelson and Robert M. Solow (1958), *Linear Programming and Economic Analysis*, New York: McGraw-Hill.

Dosi, Giovanni (1982), 'Technological paradigms and technological trajectories: A suggested interpretation of the determinants and directions of technical change', *Research Policy* **11** (3), 147–62.

Dosi, Giovanni (1988), 'Sources, procedures and microeconomic effects of innovation', *Journal of Economic Literature* **26** (3), 1120–71.

Dosi, Giovanni, Christopher Freeman, Richard Nelson, Gerald Silverberg and Luc Soete (eds) (1988), *Technical Change and Economic Theory*, New York: Francis Pinter.

Edgeworth, F. Y. (1925), *Papers Relating to Political Economy*, 3 vols, Macmillan: London.

Eliasson, Gunnar (1988), 'Schumpeterian innovation, market structure, and the stability of industrial development', in Horst Hanusch (ed.), *Evolutionary Economics: Applications of Schumpeter's Ideas*, New York: Cambridge University Press.

Fabricant, Solomon (1954), 'Economic progress and economic change', in National Bureau of Economic Research, *34th Annual Report*, New York.

Fehr, Ernst, S. Gaechter and G. Kirksteiger (1997), 'Reciprocity as a contractual enforcement device', *Econometrica*, **65**, 833–60.

Fisher, Irving (1926), *Mathematical Investigations into the Theory of Value and Prices*, New Haven: Yale University Press.

Freeman, Christopher (1983), *The Long Wave and the World Economy*, Boston MA: Butterworths.

Frey, Bruno and A. Stutzer (2002), *Happiness and Economics: How the Economy and Institutions affect Well-being*, Princeton NJ: Princeton University Press.

Friedman, Milton (1953), 'The methodology of positive economics', in Milton Friedman (ed.), *Essays in Positive Economics*, Chicago IL.: University of Chicago Press.

Gabel, H. Landis and B. Sinclair-Desgagnes (1995), 'Corporate responses to environmental concerns', in Folmer, H. and H. Opschoor (eds), *Principles of Environmental and Resource Economics*, Cheltenham, UK and Lyme MA: Edward Elgar Publishing.

Gabel, H. Landis and B. Sinclair-Desgagnes (1998), 'The firm, its routines and the environment', in T. Tietenberg and H. Folmer (eds), *The International Yearbook of Environmental and Resource Economics 1998/1999: A Survey of Current Issues*, Cheltenham, UK and Lyme MA: Edward Elgar Publishing.

Georgescu-Roegen, Nicholas (1966), *Analytic Economics*, Cambridge MA: Harvard University Press.

Georgescu-Roegen, Nicholas (1971), *The Entropy Law and the Economic Process*, Cambridge MA: Harvard University Press.

Georgescu-Roegen, Nicholas (1976), 'The economics of production', in *Energy and Economic Myths: Institutional and Analytic Economic Essays*, New York: Pergamon Press.

Georgescu-Roegen, Nicholas (1977), 'The steady state and ecological salvation: A thermodynamic analysis', *BioScience* **27** (4).

Georgescu-Roegen, Nicholas (1979), 'Energy analysis and economic valuation', *Southern Economic Journal*, **45**, 1023–58.

Georgescu-Roegen, Nicholas (1984), 'Feasible recipes and viable technologies', *Atlantic Economic Journal* 12, 21–30.

Gintis, Herbert (2000a), 'Biology meets economics', in *Game Theory Evolving*, Princeton NJ: Princeton University Press.

Gintis, Herbert (2000b), *Game Theory Evolving*, Princeton NJ: Princeton University Press.

Glansdorff, P. and Ilya Prigogine (1971), *Thermodynamic Theory of Structure, Stability and Fluctuations*, New York: Wiley-Interscience.

Goeller, H. and Alvin Weinberg (1976), 'The age of substitutability', *Science* **191**.

Guha, Ashok (1981), *An Evolutionary View of Economic Growth*, Oxford, UK: Oxford University Press.

Hanusch, Horst (ed.) (1988), *Evolutionary Economics: Applications of Schumpeter's Ideas*, New York: Cambridge University Press.

Harcourt, G. C. (1972), *Some Cambridge Controversies in the Theory of Capital*, Cambridge, UK: Cambridge University Press.

Hardin, G. (1968), 'The tragedy of the commons', *Science* **162**, 1243–8.

Harsanyi, John C. (1962), 'Bargaining in ignorance of the opponents' utility functions', *Journal of Conflict Resolution*.

Harsanyi, John C. (1966), 'A general theory of rational behavior in game situations', *Econometrica* 34, 613–34.

Hartley, K. (1969), 'Estimating military aircraft production outlays: The British experience', *Economic Journal* **79**, 861–81.

Helmstaedter, E. and M. Perlman (1996), *Behavioral Norms, Technological*

Progress and Economic Dynamics: Studies in Schumpeterian Economics, Ann Arbor MI: University of Michigan Press.

Henrich, J., R. Boyd, S. Bowles, C. Camerer, E. Fehr, Herbert Gintis and R. McElreath (2001), 'In search of Homo Economicus: Behavioral experiments in 15 small-scale societies', *American Economic Review* **91** (May), 73–8.

Hobsbawm, Eric (1975), *The Age of Capital 1848–1875*, 4th impression, 1997 edn, London: Weidenfeld & Nicolson.

Holland, John H. (1986), *Evolution, Games and Learning*, A. Lapedes, J. D. Farmer, N. H. Packard and B. Wendroff (eds), Amsterdam: North Holland.

Holland, John H. (1988), 'The dynamics of searches directed by genetic algorithms', in Y. C. Lee (ed.), *Evolution, Learning and Cognition*, Singapore: World Scientific.

Hotelling, H. (1931), 'The economics of exhaustible resource', *Journal of Political Economy* 39, 137–75.

Jackson, Tim, and Nick Marks (1994), *Measuring Sustainable Economic Welfare: A Pilot Index: 1950–1990*, Stockholm Environmental Institute report, Stockholm.

Jacobs, Jane (1992), *Systems of Survival*, New York: Random House.

Jensen, Michael C. and William H. Meckling (1994), 'The nature of man', *Journal of Applied Corporate Finance* **7** (2), 4–19.

Jewkes, John, David Sawers and Richard Stillerman (1958), *The Sources of Invention*, London: Macmillan.

Johnson, W. (1913), 'The pure theory of utility curves', *Economic Journal* **23**, 483–513.

Jorgenson, Dale W. and Kevin J. Stiroh (1995), 'Computers and growth', *Economics of Invention and New Technology* **3**, 295–316.

Jorgenson, Dale W. and Kevin J. Stiroh (2000), 'Raising the speed limit: US economic growth in the information age', *Brookings Papers on Economic Activity* **1**, 125–211.

Kahneman, D. and A. Tversky (1982), 'The simulation heuristic', in D. Kahneman et al (eds), *Judgment Under Uncertainty: Heuristic and Biases*, Cambridge: Cambridge University Press.

Kahneman, Daniel, P. Wakker and R. Sarin (1997), 'Back to Bentham? Explorations of experienced utility', *Quarterly Journal of Economics* **112**, 375–405.

Kaldor, Niko (1971), *Economics without Equilibrium*, Armonk NY: M. E. Sharpe Inc.

Kimura, Motoo (1979), 'The neutral theory of molecular evolution', *Scientific American*.

Kleinknecht, Alfred (1987), *Are there Schumpeterian Waves of Innovations?*

International Institute for Applied Systems Analysis working paper WP-87-076, Laxenburg, Austria.

Kondratieff, N. D. (1926), 'Die langen Wellen der Konjunktur', *Archiv für Sozialwissenschaft und Sozialpolitik* **56**, 573.

Koning, N., and R. Jongeneel (1997), 'Neo-Paretian welfare economics: Misconceptions and abuses', in *Wageningen University economic papers, 05-97*, Wageningen, Netherlands.

Kornai, Janos (1973), *Anti-equilibrium*, New York: North Holland Publishing Company.

Kwasnicki, Witold (1996), 'Technological development: An evolutionary model and case study', *Technological Forecasting and Social Change* **52**, 31–57.

Lancaster, Kelvin (1971), *Consumer Demand: A New Approach*, vol 5, New York: Columbia University Press.

Lane, R. (2000), *The Loss of Happiness in Market Economies*, New Haven and London: Yale University Press.

Layard, R. (2003), 'Happiness: Has social science got a clue?' paper read at Lionel Robbins Memorial Lecture Series at London School of Economics, March 3–5.

Lorenz, H. W. (1989), *Nonlinear Dynamical Motion and Chaotic Motion*, Berlin: Springer-Verlag.

Lucas, Robert E. (1976), 'Econometric policy evaluations; a critique', *Journal of Monetary Economics* (supplement), 19–46.

Lucas, Robert E., Jr. (1988), 'On the mechanics of economic development', *Journal of Monetary Economics* **22** (1), 2–42.

Maddison, Angus (1995), *Explaining the Economic Performance of Nations*, Cambridge, UK: Cambridge University Press.

Mankiw, N. Gregory (1997), *Macroeconomics*, New York: Worth Publishing.

Mansfield, Edwin (1965), 'Rates of return from industrial R and D', *American Economic Review* **55** (2), 310–22.

Mansfield, Edwin (1981), 'Composition of R and D expenditures: Relationship to size of firm, concentration and innovative output', *Review of Economics and Statistics*, 610–14.

Mansfield, Edwin (1983), 'Long waves and technological innovation', *American Economic Review* **73** (2), 141–5.

Mansfield, Edwin et al. (1977), 'Social and private rates of return from industrial innovation', *Quarterly Journal of Economics* **91** (2), 221–40.

Mansfield, Edwin et al. (1983), 'New findings in technology transfer, productivity and economic policy', *Research Management*, 11–20.

Marshall, Alfred (1977), *Principles of Economics*, 8th reprint, reset of 1890 edn in 5 vols, vol V, London: Macmillan.

Martinás, Katalin (1989), *About Irreversibility in Microeconomics*, Roland Eotvos University Department of Low Temperature Physics research report AHFT-89-1, March, Budapest, Hungary.

Martinás, Katalin (1996), 'Irreversible microeconomics', paper read at Workshop on Methods of Non-equilibrium Processes, Budapest, Hungary.

Martinás, Katalin (2000), 'About irreversibility in economics', *Open Systems Information Dynamics* 7, 349–64.

Martinás, Katalin (2003), 'Is the utility maximum principle necessary?' in E. Fullbrook (ed.), *Crisis in Economics*, London: Routledge.

Marx, Karl (1867), *Das Kapital*, German edn. 2 vols.

Maslow, Abraham H. (1943), 'A theory of human motivation', *Psychological Review* 50 (January), 370–96.

Maslow, Abraham H. (1970), *Motivation and Personality*, 2nd edn, New York: Harper and Row.

Meadows, Donella H., Dennis L. Meadows, Jorgen Randers and William W. Behrens III (1972), *The Limits to Growth: A Report for the Club of Rome's Project on the Predicament of Mankind*, New York: Universe Books.

Menger, Carl (1981), *Principles of Economics*, translated by J. Dingwell and B. F. Hoselitz, New York: New York University Press.

Mirowski, Philip (1984), 'The role of conservation principles in twentieth century economic theory', *Philosophy of the Social Sciences* 14, 461–73.

Mirowski, Philip (1989a), 'More heat than light: Economics as social physics; physics as nature's economics', in paperback edn, Craufurd D. Goodwin (ed.), *Historical Perspectives on Modern Economics*, Cambridge, UK: Cambridge University Press.

Mirowski, Philip (1989b), 'On Hollander's substantive identity of classical and neoclassical economics: A reply', *Cambridge Journal of Economics* 13, 471–7.

Nash, John F. (1951), 'Non-cooperative games', *Annals of Mathematics* 54, 286–95.

Nash, John F. (1953), 'Two person cooperative games', *Econometrica* 21, 128–40.

Nelson, Richard P. (ed.) (1962), *The Rate and Direction of Inventive Activity*, Princeton NJ: Princeton University Press.

Nelson, Richard R. (1959), 'The simple economics of basic scientific research', *Journal of Political Economy* 67, 297–306.

Nelson, Richard R. (1982), 'The role of knowledge in R and D efficiency', *Quarterly Journal of Economics* 97, 453–70.

Nelson, Richard R. and Sidney G. Winter (1974), 'Neoclassical vs. evolutionary theories of economic growth: Critique and prospectus', *Economic Journal* 84 (336), 886–905.

Nelson, Richard R. and Sidney G. Winter (1982), *An Evolutionary Theory of Economic Change*, Cambridge MA: Harvard University Press.

Newbery, David and Joseph Stiglitz (1982), 'The choice of techniques and the optimality of market equilibrium with rational expectations', *Journal of Political Economy* **90**.

Ng, Yew-Kwang (1997), 'A case for happiness, cardinalism, and interpersonal comparability', *Economic Journal* **107**, 1848–58.

Nicolis, Gregoire and Ilya Prigogine (1977), *Self-organization in Non-equilibrium Systems*, New York: Wiley-Interscience.

Nordhaus, William D. (2004), 'Schumpeterian profits in the American economy: Theory and measurement', Cowles Foundation for Research in Economics, discussion paper: 1457, Yale University, New Haven, CT, 28 April.

Pearce, David W. and Giles Atkinson (1993), 'Capital theory and the measurement of sustainable development: An indicator of weak sustainability', *Ecological Economics* **8**, 166–81.

Pearce, David W., K. Hamilton and Giles Atkinson (1996), 'Measuring sustainable development: Progress on indicators', *Environment and Development Economics* **1**, 85–101.

Perez-Perez, Carlotta (1983), 'Towards a comprehensive theory of long waves', in G. Bianchi, G. Bruckman, J. Delbeke and T. Vasko (eds), *Long Waves, Depression and Innovation*, proceedings of the IIASA Siena/Florence meeting, Laxenburg, Austria, 26–30 October.

Perrings, Charles (1997), 'Georgescu-Roegen and the irreversibility of material processes', *Ecological Economics* **22** (3), 303–4.

Pezzey, John (1989), *Economic Analysis of Sustainable Growth and Sustainable Development*, World Bank. Environment Department working paper 15, Washington DC.

Pigou, A. C. (1952), *The Economics of Welfare*, 2nd edn, London: Macmillan Company.

Pontryagin, Lev Semenovich (1962), *Mathematical Theory of Optimal Processes*.

Prigogine, Ilya, Gregoire Nicolis and A. Babloyantz (1972), 'Thermodynamics of evolution', *Physics Today* **23** (11/12), 23–28(N) and 38–44(D).

Radner, Roy (1968), 'Competitive equilibrium under uncertainty', *Econometrica* **36**, 31–58.

Rapping, Leonard (1965), 'Learning and world war II production functions', *Review of Economics and Statistics* **47**, 81.

Robinson, Joan (1955a), 'The production function', *Economic Journal* **65** (257), 67.

Robinson, Joan (1955b), 'The production function and the theory of capital', *Review of Economic Studies* **55**, 81ff.

Romer, Paul M. (1986), 'Increasing returns and long-run growth', *Journal of Political Economy* **94** (5), 1002–37.

Romer, Paul M. (1987), 'Growth based on increasing returns due to specialization', *American Economic Review* **77** (2), 56–62.

Romer, Paul M. (1990), 'Endogenous technological change', *Journal of Political Economy* **98** (5), S71–S102.

Rosen, S. (1972), 'Learning by experience as joint production', *Quarterly Journal of Economics* **86**, 366–82.

Rosenberg, Nathan (1969), 'The direction of technological change: Inducement mechanisms and focusing devices', *Economic Development and Cultural Change* **18**, 1–24.

Rosenberg, Nathan (1976), *Perspectives in Technology*, New York: Cambridge University Press.

Rosenberg, Nathan (1982), *Inside the Black Box: Technology and Economics*, New York: Cambridge University Press.

Rosenberg, Nathan and Claudio R. Frischtak (1984), 'Technological innovation and long waves', *Cambridge Journal of Economics* **8**, 7–24.

Ross, S. (1973), 'The economic theory of agency', *American Economic Review* **63**, 134–9.

Rostow, W. W. (1975), 'Kondratieff, Schumpeter, Kuznets: Trend periods revisited', *Journal of Economic History* **35**.

Samuelson, Paul (1954), 'The pure theory of public expenditure', *Review of Economics and Statistics*, 387–9.

Sargent, T. J. and L. P. Hansen (1981), 'Linear rational expectations models of dynamically interrelated variables', in *Rational Expectations and Econometric Practice*, Minneapolis, MN: University of Minnesota Press.

Schumpeter, Joseph A. (1912), *Theorie der Wirtschaftlichen Entwicklungen*, Leipzig, Germany: Duncker and Humboldt.

Schumpeter, Joseph A. (1934), *Theory of Economic Development*, Cambridge MA: Harvard University Press.

Schwefel, Hans-Paul (1988), *Evolutionary Learning Optimum-seeking on Parallel Computer Architectures*, Dortmund, Germany: University of Dortmund Department of Computer Science.

Scitovsky, T. (1954), 'Two concepts of external economies', *Journal of Political Economy* **62**, 143–51.

Shoemaker, Paul (1982), 'The expected utility model', *Journal of Economic Literature* **20**, 528–63.

Shubik, Martin (1982), *Game Theory in the Social Sciences*, Cambridge MA: MIT Press.

Silverberg, Gerald (1988), 'Modeling economic dynamics and technical change: Mathematical approaches to self-organization and evolution', in Giovanni Dosi, Christopher Freeman, Richard Nelson, Gerald Silverberg and Luc Soete (eds), *Technical Change and Economic Theory*, London: Pinter Publishers.

Silverberg, Gerald, Giovanni Dosi and Luigi Orsenigo (1988), 'Innovation, diversity and diffusion: A self-organizing model', *Economic Journal* **98** (393), 1032–54.

Silverberg, Gerald and D. Lehnert (1993), 'Long waves and evolutionary chaos in a simple Schumpeterian model of embodied technological change', *Structural Change & Economic Dynamics* **4**, 9–37.

Silverberg, Gerald and B. Verspagen (1994a), 'Collective learning, innovation and growth in a boundedly rational, evolutionary world', *Journal of Evolutionary Economics* **4**:207–26.

Silverberg, Gerald and B. Verspagen (1994b), 'Learning, innovation and economic growth: A long-run model of industrial dynamics', *Industrial and Corporate Change* **3**, 199–223.

Silverberg, Gerald and B. Verspagen (1996), 'From the artificial to the endogenous: Modeling evolutionary adaptation and economic growth', in E. Helmstaedter and M. Perlman (eds), *Behavioral Norms, Technological Progress and Economic Dynamics: Studies in Schumpeterian Economics*, Ann Arbor MI: University of Michigan Press.

Simon, Herbert A. (1955), 'A behavioral model of rational choice', *Quarterly Journal of Economics* **69**, 99–118.

Simon, Herbert A. (1959), 'Theories of decision-making in economics', *American Economic Review* **49**, 253–83.

Smith, Adam (1776 [1976]), 'An inquiry into the nature and causes of the wealth of nations', in *Collected Works of Adam Smith*, Oxford, UK: Clarendon Press.

Smith, V. Kerry and John Krutilla (eds) (1979), *Scarcity and Growth Revisited*, Baltimore, MD: Johns Hopkins University Press.

Solow, Robert M. (1956), 'A contribution to the theory of economic growth', *Quarterly Journal of Economics* **70**, 65–94.

Solow, Robert M. (1957), 'Technical change and the aggregate production function', *Review of Economics and Statistics* **39**, 312–20.

Solow, Robert M. (1970), 'Foreword', in Edwin Burmeister and A. Rodney Dobell (eds), *Mathematical Theories of Economic Growth*, New York: Macmillan Company.

Solow, Robert M. (1974), 'The economics of resources or the resources of economics', *American Economic Review* **64**.

Solow, Robert M. (1986), 'On the intergenerational allocation of natural resources', *Scandinavian Journal of Economics* **88**, 141–9.

Solow, Robert M. (1992), *An Almost Practical Step towards Sustainability*, Washington DC: Resources for the Future.

Stern, David I. (1997), 'Limits to substitution and irreversibility in production and consumption: a neoclassical interpretation of ecological economics', *Ecological Economics* **22**, 197–215.

Stiglitz, Joseph (1974), 'Growth with exhaustible natural resources: Efficient and optimal growth paths', *Review of Economic Studies*.

Stokey, Nancy L. (1986), 'The dynamics of industrywide learning', in *Social Choice and Public Decision Making: Essays in Honor of Kenneth J. Arrow*, New York: Cambridge University Press.

Suzumura, K. (1999), 'Paretian welfare judgments and Bersonian social choice', *Economic Journal* **109** (April), 204–21.

Takase, Kae and Yasuhiro Murota (2004), 'The impact of IT investment on energy: Japan and US comparison in 2010', *Energy Policy* **32**, 1291–301.

Tobin, James and William Nordhaus (1972), 'Is growth obsolete?' in *Economic Growth*, New York: Columbia University Press.

Toman, Michael A., John Pezzey and J. Krautkraemer (1995), 'Neoclassical economic growth theory and sustainability', in D. W. Bromley (ed.), *Handbook of Environmental Economics*, Oxford, UK: Blackwell Publishers.

Tversky, A. and D. Kahneman (1974), 'Judgment under uncertainty: Heuristics and biases', *Science* **185**, 1124–31.

Tversky, A. and D. Kahneman (1981), 'The framing of decisions and the psychology of choice', *Science* **211**, 453–8.

Tversky, A. and D. Kahneman (1987), 'Rational choice and the framing of decisions', in R. M. Hogarth and M. W. Reder (eds), *Rational Choice: The Contrast Between Economics and Psychology*, Chicago: University of Chicago Press.

van den Bergh, Jeroen C. J. M., Ada Ferrer-i-Carbonell, and Giuseppe Munda (2000), 'Alternative models of individual behavior and implications for environmental policy', *Ecological Economics* **32**, 43–61.

van der Zwan, A. (1979), *On the Assessment of the Kondratieff Cycle and Related Issues*, Rotterdam, The Netherlands: Center for Research and Business.

von Neumann, John (1945), 'A model of general economic equilibrium', *Review of Economic Studies* **13**, 1–9.

von Neumann, John and Oskar Morgenstern (1944), *Theory of Games and Economic Behavior*, Princeton NJ: Princeton University Press.

Vromen, Jack J. (1995), *Economic Evolution*, London: Routledge.

Waller, John (2004), *Fabulous Science: Fact and Fiction in the History of Scientific Discovery*, first paperback edn, Oxford: Oxford University Press.

Weatherford, Jack (1997), *The History of Money*, Paperback edn, New York: Three Rivers Press.

Wicksteed, Philip Henry (1910), *The Common Sense of Political Economy, including a Study of the Human Basis of Economic Law*, London: Macmillan and Co., Limited.

Wiener, Norbert (1948), *Cybernetics: Control and Communications in the Animal and the Machine*, New York: John Wiley and Sons.

Winter, Sidney G. (1964), 'Economic "natural selection" and the theory of the firm', *Yale Economic Essays* **4** (1), 225–72.

Winter, Sidney G. (1984), 'Schumpeterian competition in alternative technological regimes', *Journal of Economic Behavior and Organization* **5** (3–4), 287–320.

Wright, T. P. (1936), 'Factors affecting the cost of airplanes', *Journal of Aeronautical Sciences* **3**, 122–8.

Yergin, Daniel (1991), *The Prize: The Epic Quest for Oil, Money and Power*, New York: Simon and Schuster.

Index